# Biomass—A Renewable Resource for Carbon Materials

# Biomass—A Renewable Resource for Carbon Materials

Editors

**Indra Neel Pulidindi**
**Pankaj Sharma**
**Aharon Gedanken**

MDPI • Basel • Beijing • Wuhan • Barcelona • Belgrade • Manchester • Tokyo • Cluj • Tianjin

*Editors*

Indra Neel Pulidindi
School of Science
GSFC Univeristy
Vadodara
India

Pankaj Sharma
Department of Applied
Chemistry
The Maharaja Sayajirao
University of Baroda
Vadodara
India

Aharon Gedanken
Department of Chemistry
Bar Ilan University
Ramat Gan
Israel

*Editorial Office*
MDPI
St. Alban-Anlage 66
4052 Basel, Switzerland

This is a reprint of articles from the Special Issue published online in the open access journal *C* (ISSN 2311-5629) (available at: www.mdpi.com/journal/carbon/special_issues/ZC6T264W57).

For citation purposes, cite each article independently as indicated on the article page online and as indicated below:

LastName, A.A.; LastName, B.B.; LastName, C.C. Article Title. *Journal Name* **Year**, *Volume Number*, Page Range.

**ISBN 978-3-0365-7189-8 (Hbk)**
**ISBN 978-3-0365-7188-1 (PDF)**

© 2023 by the authors. Articles in this book are Open Access and distributed under the Creative Commons Attribution (CC BY) license, which allows users to download, copy and build upon published articles, as long as the author and publisher are properly credited, which ensures maximum dissemination and a wider impact of our publications.

The book as a whole is distributed by MDPI under the terms and conditions of the Creative Commons license CC BY-NC-ND.

# Contents

About the Editors . . . . . . . . . . . . . . . . . . . . . . . . . . . . . . . . . . . . . . . . . . . . . . . . . . . . vii

Preface to "Biomass—A Renewable Resource for Carbon Materials" . . . . . . . . . . . . . . . ix

**Otmane Sarti, Fouad El Mansouri, Emilia Otal, José Morillo, Abdelhamid Ouassini and Jamal Brigui et al.**
Assessing the Effect of Intensive Agriculture and Sandy Soil Properties on Groundwater Contamination by Nitrate and Potential Improvement Using Olive Pomace Biomass Slag (OPBS)
Reprinted from: C **2022**, 9, 1, doi:10.3390/c9010001 . . . . . . . . . . . . . . . . . . . . . . . . . . . . 1

**Arcadius Martinien Agassin Ahogle, Felix Kouelo Alladassi, Tobi Moriaque Akplo, Hessou Anastase Azontonde and Pascal Houngnandan**
Assessing Soil Organic Carbon Stocks and Particle-Size Fractions across Cropping Systems in the Kiti Sub-Watershed in Central Benin
Reprinted from: C **2022**, 8, 67, doi:10.3390/c8040067 . . . . . . . . . . . . . . . . . . . . . . . . . . . 23

**Dinesh Bejjanki, Praveen Banothu, Vijay Bhooshan Kumar and Puttapati Sampath Kumar**
Biomass-Derived N-Doped Activated Carbon from Eucalyptus Leaves as an Efficient Supercapacitor Electrode Material
Reprinted from: C **2023**, 9, 24, doi:10.3390/c9010024 . . . . . . . . . . . . . . . . . . . . . . . . . . . 39

**Abdualilah Albaiz, Muhammad Alsaidan, Abdullah Alzahrani, Hassan Almoalim, Ali Rinaldi and Almaz S. Jalilov**
Active Carbon-Based Electrode Materials from Petroleum Waste for Supercapacitors
Reprinted from: C **2022**, 9, 4, doi:10.3390/c9010004 . . . . . . . . . . . . . . . . . . . . . . . . . . . 51

**Najla Grioui, Amal Elleuch, Kamel Halouani and Yongdan Li**
Valorization of Exhausted Olive Pomace for the Production of a Fuel for Direct Carbon Fuel Cell
Reprinted from: C **2023**, 9, 22, doi:10.3390/c9010022 . . . . . . . . . . . . . . . . . . . . . . . . . . . 63

**Luís Carmo-Calado, Roberta Mota-Panizio, Ana Carolina Assis, Catarina Nobre, Octávio Alves and Gonçalo Lourinho et al.**
Pre-Feasibility Study of a Multi-Product Biorefinery for the Production of Essential Oils and Biomethane
Reprinted from: C **2022**, 9, 2, doi:10.3390/c9010002 . . . . . . . . . . . . . . . . . . . . . . . . . . . 89

**Tetiana Mirzoieva, Nazar Tkach, Vitalii Nitsenko, Nataliia Gerasymchuk, Olga Tomashevska and Oleksandr Nechyporenko**
Development of the Correlation Model between Biogas Yield and Types of Organic Mass and Analysis of Its Key Factors
Reprinted from: C **2022**, 8, 73, doi:10.3390/c8040073 . . . . . . . . . . . . . . . . . . . . . . . . . . . 103

**Nesrine Madani, Imane Moulefera, Souad Boumad, Diego Cazorla-Amorós, Francisco José Varela Gandía and Ouiza Cherifi et al.**
Activated Carbon from *Stipa tenacissima* for the Adsorption of Atenolol
Reprinted from: C **2022**, 8, 66, doi:10.3390/c8040066 . . . . . . . . . . . . . . . . . . . . . . . . . . . 119

**Denis Miroshnichenko, Katerina Lebedeva, Anna Cherkashina, Vladimir Lebedev, Oleksandr Tsereniuk and Natalia Krygina**
Study of Hybrid Modification with Humic Acids of Environmentally Safe Biodegradable Hydrogel Films Based on Hydroxypropyl Methylcellulose
Reprinted from: C **2022**, 8, 71, doi:10.3390/c8040071 . . . . . . . . . . . . . . . . . . . . . . . . . . . 135

**Steven C. Peterson and A. J. Thomas**
Lauric Acid Treatments to Oxidized and Control Biochars and Their Effects on Rubber Composite Tensile Properties
Reprinted from: C **2022**, *8*, 58, doi:10.3390/c8040058 . . . . . . . . . . . . . . . . . . . . . . . . **145**

# About the Editors

**Indra Neel Pulidindi**

Dr Indra Neel Pulidindi is currently (from 19$^{th}$ September 2022 to till date) working as an assistant professor in GSFC University, India. He received his PhD degree from the Indian Institute of Technology Madras in the year 2010 under the supervision of Professor T K Varadarajan and Professor (Em) B Viswanathan. He worked in the laboratory of Professor Aharon Gedanken from 2010 to 2016 in Israel and worked on biomass conversion of biofuels and biochemicals. From 2016–2017, he worked in the laboratory of Professor Tae Hyun Kim at Hanyang University and made systematic studies on biomass composition analysis and conversion of biomass to biochemicals. Subsequently, he worked in the laboratory of Professor Xinling Wang at Shanghai Jiao Tong University during 2019–2021, on carbon fiber reinforced composite materials. Dr Neel has 47 research papers, 1 patent, 4 patent applications, 1 book, 1 ebook and 6 book chapters to his credit. Dr Neel guided several PhD, masters, undergraduates and freshmen in their academic research curriculum and helped them to earn their degrees. Thus, Dr Neel has been a dynamic researcher and teacher all through his career thus far.

**Pankaj Sharma**

Pankaj Sharma, Assistant Professor, Department of Applied Chemistry, Faculty of Technology and Engineering, The Maharaja Sayajirao University of Baroda, Vadodara, Gujarat, INDIA.

He has been carrying out research in multidisciplinary area of chemistry like Drug chemistry, Removal of Genotoxic Impurities from API, Energy storage materials, Mesoporous materials, Ionic Liquids, Carbon dioxide Capture and Sequestration, Green Chemistry, Heterogeneous catalysis.

Till now he has published 19 peer reviewed SCI and SCIE papers. In addition to this, he 01 Book chapter in Elsevier and he is Editor of 01 Book. He is guest editor of 04 MDPI journals. He has been granted 02 patents also, one Korean and one US patent. His h index is 12 and i 10 index is 11.

He evaluated 03 Ph. D. thesis as Indian Referee and took the Final Ph. D defense Viva Voce of 02 students. He is guiding 01 Ph. D student and 04 MSc. Final students for their dissertation project. 20 MSc. Final students have completed their dissertation project under him.

Before Ph.D., he acquired the skills of a researcher by working in the India's most advanced CSIR group of laboratories like Central Drug Research Institute (CDRI) and Central Salt and Marine Chemicals Research Institute (CSMCRI). But his research skills achieved a new height by working with his Korean mentors, Dr. Il Hyun Baek at Korea Institute of Energy Research (KIER), Prof. Chang Sik Ha at Pusan National University (PNU) and Prof. Jae Won Lee at Dankook University (DU) in Republic of Korea respectively in their well established and highly advanced laboratory during his postdoctoral experience.

**Aharon Gedanken**

Professor Aharon Gedanken is an outstanding and legendary scientist with over 850 publications in peer reviewed journals of international repute and with high impact factors. He has 37 patent applications and a book on biofuels and over 10 book chapters to his credit. Professor Gedanken made remarkable contributions to the fields of sonochemistry and microwave technology and their application to nanomaterials, biomaterials and biofuels. He served as faculty in the Department of Chemistry at Bar Ilan University for over 34 years (1975–2009) and has been an Emeritus professor in the same institute for over 13 years (2009–to date). Professor Gedanken's research metrics, namely, H-index and citations, are 117 and 51510, respectively. His research interests include Solid State Chemistry, Catalysis, Energy, Materials Science, and Biochemistry.

# Preface to "Biomass—A Renewable Resource for Carbon Materials"

Currently, humanity is indeed in turbulent times. The current world affairs are no different from those during World War I (28$^{th}$ July 1914–11$^{th}$ November 1918) or during the time of World War II (1$^{st}$ September 1939–2$^{nd}$ September 1945). Scientists, especially, chemists, have a vital role to ensure that lifesaving materials are available at the disposal of common people. Renowned German chemist Raphael von Ostrejko has a few patents based on activated carbon from biomass during 1900–1920 and his inventions based on those patents, namely activated carbon based masks, saved many people from perishing due to the harmful effects of poisonous gases used during World War I. Carbon materials from biomass, in particular, are equally significant during these testing times of war, energy and food shortages and climate change. Biomass constitutes a rich source of carbon materials with diverse properties and applications impacting almost every sphere of human activity, ranging from agriculture, health, energy, environment, materials, safety, security, defence and many more. All the three vital components of biomass, namely cellulose, hemicellulose and lignin, offer unique structure and morphological features to the resulting carbon materials. Owing to the sustainability, diversity in properties and applications of these carbon materials produced from biomass, it is conceived that a Special Issue needed to be launched in the journal "C-journal of carbon research" published by Multidisciplinary Digital Publishing Institute (MDPI), wherein the latest scientific advancements throughout the globe could be assembled and made available freely, at a click, to the scientific, academic and industrial fraternity for their growth and development. Thus, this Special Issue was launched with the title "Biomass—A renewable resource for carbon materials". Indeed, the attempt has been a great success due to the untiring and visionary approach of the managing editor Dr William Wang. Within a short duration of 4.5 months (4/8/2022 to 20/12/2022), we succeeded in publishing 10 state of the art research papers on diverse applications of carbon materials, namely, agriculture, energy, environment and medicine. The editors of this Special Issue place on record their indebtedness to the 10 research groups that have contributed 10 outstanding papers to the Special Issue, facilitating the conversion of the Special Issue to a published reference book on carbon materials with 10 chapters.

Finally, I bow down before my LORD and Saviour Jesus Christ for His all sufficient grace that enabled His servant to succeed in this scientific pursuit.

Dedicated to My Lord and My God Jesus Christ. My grace is sufficient for thee.
(2 Corinthians 12:9).

**Indra Neel Pulidindi, Pankaj Sharma, and Aharon Gedanken**
*Editors*

Article

# Assessing the Effect of Intensive Agriculture and Sandy Soil Properties on Groundwater Contamination by Nitrate and Potential Improvement Using Olive Pomace Biomass Slag (OPBS)

Otmane Sarti [1], Fouad El Mansouri [2,*], Emilia Otal [3], José Morillo [3], Abdelhamid Ouassini [1], Jamal Brigui [2] and Mohamed Saidi [1]

[1] Laboratory of LAMSE, Faculty of Sciences and Techniques of Tangier, B.P. 416, Tangier 90000, Morocco
[2] Research Team: Materials, Environment and Sustainable Development (MEDD), Faculty of Sciences and Techniques of Tangier, B.P. 416, Tangier 90000, Morocco
[3] Department of Chemical and Environmental Engineering, University of Seville, Camino de Los Descubrimientos, s/n, 41092 Seville, Spain
* Correspondence: fouad.elmansouri@etu.uae.ac.ma; Tel.: +212-662-102-847

**Abstract:** The relationship between agricultural activities, soil characteristics, and groundwater quality is critical, particularly in rural areas where groundwater directly supplies local people. In this paper, three agricultural sandy soils were sampled and analyzed for physicochemical parameters such as pH, water content, bulk density, electrical conductivity (EC), organic matter (OM), cation exchange capacity (CEC), and soil grain size distribution. Major and trace elements were analyzed by inductively coupled plasma-optical emission spectrometry (ICP/OES) to determine their concentrations in the fine fraction (FF) of the soils. Afterward, the elemental composition of the soils was identified by X-ray powder diffraction (XRD) and quantified by X-ray fluorescence (XRF). The surface soil characteristics were determined by the Brunauer–Emmett–Teller (BET) method, whereas the thermal decomposition of the soils was carried out using thermogravimetric analysis and differential scanning calorimetric (TGA-DSC) measurements. The morphological characteristics were obtained by scanning electron microscopy (SEM). Afterward, column-leaching experiments were conducted to investigate the soil's retention capacity of nitrate ($NO_3^-$). Parallelly, a chemical and physical study of olive pomace biomass slag (OPBS) residue was carried out in order to explore its potential use as a soil additive and improver in the R'mel area. The OPBS was characterized by physicochemical analysis, assessed for heavy metals toxicity, and characterized using (XRD, XRF, SEM, and BET) techniques. The results show that the R'mel soils were slightly acidic to alkaline in nature. The soils had a sandy texture with low clay and silt percentage (<5% of the total fraction), low OM content, and weak CEC. The column experiments demonstrated that the R'mel irrigated soils have a higher tendency to release large amounts of nitrate due to their texture and a higher degree of mineralization which allows water to drain quickly. The OPBS chemical characterization indicates a higher alkaline pH (12.1), higher water content (7.18%), and higher unburned carbon portion (19.97%). The trace elements were present in low concentrations in OPBS. Macronutrients in OPBS showed composition rich in Ca, K, and Mg which represent 10.59, 8.24, and 1.56%, respectively. Those nutrients were quite low in soil samples. Both XRD and XRF characterization have shown a quasi-dominance of $SiO_2$ in soil samples revealing that quartz was the main crystalline phase dominating the R'mel soils. Oppositely, OPBS showed a reduced $SiO_2$ percentage of 26,29% while K, Ca, and P were present in significant amounts. These results were confirmed by XRF analysis of OPBS reporting the presence of dolomite (CaMg, $(CO_3)_2$), fairchildite ($K_2Ca (CO_3)_2$), and free lime (CaO). Finally, the comparison between the surface characteristic of OPBS and soils by BET and SEM indicated that OPBS has a higher surface area and pore volume compared to soils. In this context, this study suggests a potential utilization of OPBS in order to (1) increase soil fertility by the input of organic carbon and macronutrients in soil; (2) increase the water-holding capacity of soil; (3) increase soil CEC; (4) stabilize trace elements; (5) enhance the soil adsorption capacity and porosity.

Citation: Sarti, O.; El Mansouri, F.; Otal, E.; Morillo, J.; Ouassini, A.; Brigui, J.; Saidi, M. Assessing the Effect of Intensive Agriculture and Sandy Soil Properties on Groundwater Contamination by Nitrate and Potential Improvement Using Olive Pomace Biomass Slag (OPBS). C 2023, 9, 1. https://doi.org/10.3390/c9010001

Academic Editors: Indra Pulidindi, Pankaj Sharma and Aharon Gedanken

Received: 25 October 2022
Revised: 15 December 2022
Accepted: 20 December 2022
Published: 22 December 2022

Copyright: © 2022 by the authors. Licensee MDPI, Basel, Switzerland. This article is an open access article distributed under the terms and conditions of the Creative Commons Attribution (CC BY) license (https://creativecommons.org/licenses/by/4.0/).

**Keywords:** nitrate contamination; groundwater; leaching; soil chemical characterization; biomass valorization

## 1. Introduction

Soil is one of the most essential elements in life. Its functions are crucial to the ecosystem because it is considered a storehouse of carbon and a food supplier. In addition, healthy soils are a prerequisite for ensuring the ecological ecosystem functions worldwide [1,2]. Moreover, the soil has a primordial role in limiting the intrusion of pollutants in groundwater by acting as a filter [3]. In several irrigated areas, worrying signs of deterioration in water and soil quality have been reported. Agricultural practices directly impact the soil's physical, chemical, and biological properties [4,5]. The alteration of soil properties has resulted in many environmental problems, such as soil degradation, salinization, waterlogging [6], deforestation and erosion [7], and groundwater contamination. The terms of agriculture conservation must be respected by ensuring the recycling of nutrients, avoiding environmental losses, and reducing the emission of greenhouse gases, whether at the regional or national scale [8]. Nitrogen is an essential macronutrient for healthy plant growth and high-yield production. Nevertheless, the massive use of nitrogenous fertilizers has led to some environmental problems, such as nitrate leaching [9,10]. After N application, crops assimilate their nitrogen needs by absorbing nitrate and ammonium accessible in the soil. The surplus of nitrate exceeds the plants' demand and soil denitrification capacity [11] and leaches out of the root zone as one of the most common forms of groundwater contamination [12,13]. The leaching of nitrate from the soil is a major problem threatening surface and groundwater quality and therefore human health [14,15]. The nitrate form ($NO_3^-$) of nitrogen is highly soluble, easily mobile within the soil, and poorly adsorbed by the soil particles. Recent literature shows increasing global concern about the impact of nitrate leaching with regard to the environment, especially in agricultural ecosystems [10]. The nitrate background is determined not to exceed 10 mg/L, and values exceeding this concentration indicate anthropogenic pollution [16]. The factors influencing the leaching of nitrate from the soil are numerous. Still, the most important remains the nature of the soil (the content of clay, silt, and organic matter), the irrigation and precipitation rates, the dose of fertilizers, and the temperature.

The soil texture is the most important determining factor influencing the vertical movement of contaminants through the soil. In coarse-textured sandy soils, the voids between soil particles are large in volume, allowing water to flow quickly through the unsaturated zone and reach groundwater. Huang and Hartemink [17] reported that sandy soils often have high hydraulic conductivity, gas permeability, and specific heat, but low field capacity, permanent wilting point, organic carbon, and cation exchangeable capacity. Therefore, filtration or natural water treatment takes a minimal amount of time. On the contrary, in fine-textured soils such as clays, the movement of water and contaminants through the soil is prolonged, which gives the clay minerals the time to adsorb pollutants and allows bacteria and other microorganisms to degrade contaminants before reaching the groundwater. Furthermore, the groundwater level can vary considerably from season to season, depending mainly on the infiltration rates. Consequently, the percentage of clay could be a deterministic factor affecting groundwater, especially in agricultural areas.

The soil characteristics could be determined using several characterization techniques. Simultaneous use of thermogravimetric analysis (TGA) and differential scanning calorimetry (DSC) in combination or association with XRD and other chemical analyses could be used for the quantitative determination of a particular mineral or the estimation of the total mineralogical composition [18]. Indeed, the knowledge of soil characteristics using different techniques (SEM, XRF, XRD, BET, TGA/DSC) allows the determination of the soil texture and its influence on the mobility, adsorption, and leaching rates of pollutants. The study of soil characteristics in agricultural areas could help decision-makers and scientists in understanding the processes that might reduce groundwater pollution. Parallelly, proposing low-cost solutions for soil remediation and optimization might be beneficial to the environment, especially in sandy soils. Many approaches and strategies are already in place to address soil pollution issues. Soil remediation techniques and their applicability (e.g., in situ or ex situ) differ according to the type of contamination, the method of treatment (physical, chemical, or bioremediation), and the cost-effectiveness of treatment [19]. Nanomaterial is a novel technology that is quickly evolving and expanding its domains of application in all areas of research [20,21]. Yaqoob et al. [22] noted that nanoparticles have become the most appealing and widely employed materials for a wide range of applications including agriculture and wastewater treatment. Nowadays, nanotechnology has the potential to offer solutions for agricultural challenges such as boosting nutrient utilization efficiency, mitigating heavy metal toxicity, and efficiently improving soil fertility [23,24]. Alessandrino et al. [25] investigated the ability of graphene to reduce the concentration of nitrate in sandy soils and concluded that, unlike other soil improvers, graphene can stimulate the denitrification process in soil. The use of biochar in reducing soil contamination has been extensively studied during the last years [26–29]. Due to their higher cation exchange capacity, complexation, precipitation, physisorption, and electrostatic interaction, alkaline substances such as cement, lime, fly ash, steel slag, and blast furnace slag are excellent stabilizers for soil contaminants [30,31]. Das et al. [32] highlighted the importance of reusing steel slag (steel processing by-products) to increase crop productivity and soil fertility, reduce greenhouse gas emissions, and stabilize heavy metals in contaminated soils. Liyun et al. [33] reported that steel slag is efficient for nitrate removal and might be used to decrease nitrate leaching from the soil. The fast growth of biomass power plants has resulted in massive amounts of ashes and slags [34]. The application of biomass ash and slag to agricultural soils is now largely recognized as the most efficient way for recycling these residues [35].

The groundwater resources in the Loukkos region are well known for their low quality resulting mainly from intensive farming activities. The sandy nature of the soil, the intensive use of fertilizers, and the shallow aquifer make the R'mel groundwater sensitive to physicochemical contamination. This vulnerability becomes more problematic as long as the area provides water intended for human consumption. According to Tanji et al. [36], in the same study area, 25 groundnut farmers used extensively six nitrogenous fertilizers with an average of 350 Kg/ha. Such excessive nitrogenous applications are unacceptable since the majority of these effluents would immediately infiltrate groundwater. Moreover, previous studies have reported the contamination of public drinking wells by higher concentrations of nitrate and pesticide residues in this region [37–39]. Contrariwise, no studies aim to explain the influence of soil properties on groundwater contamination by nitrate in this perimeter.

In this study, agricultural sandy soils were analyzed for physicochemical parameters and characterized in order to investigate the influence of soil properties and intensive farming on nitrate leaching and to determine the nitrate-retention capacity in sandy soils through column experiments. In addition, the authors proposed the potential utilization of biomass slag (BS) formed during the combustion of olive pomace as a soil additive and improver. To explore the physical and chemical properties of this residual material, the olive pomace biomass slag (OPBS) was evaluated for physicochemical parameters and heavy metal toxicity and characterized using different techniques (ICP/OES, XRD, XRF, BET, and SEM).

## 2. Materials and Methods

### 2.1. Presentation of the Study Area

The Loukkos perimeter is located in the northwest of Morocco between Rabat, the capital, and Tangier. The Loukkos perimeter covers an area of 256,000 Ha, with a large-scale irrigated part of 27,000 Ha. It includes the alluvial plains, located at different altitudes, the plateaus, and the roughly tortuous hills. The study area can be classified as the

Mediterranean climate, characterized by a sub-humid and temperate winter and by a hot and dry summer.

The surface water resources of the region are dominated by the Lokkous River, the most important river in the area. The hydrological regime of the Lokkous is pluvial and characterized by a strong inter-annual irregularity. A strong irregularity marks the seasonal distribution of the inputs during January and February. The villages and urban centers Ksar el Kébir and Larache are experiencing rapid growth caused by the success of the development of irrigated areas. This growth has generated an increase in various industries. However, the socio-economic infrastructure is not developing quickly due to inadequate planning and financial and human resources. This lack has generated negative repercussions on the environment of the region. It has led to the spread of unsanitary housing, especially in the surrounding rural areas, which poses health and environmental problems.

*2.2. The R'mel Groundwater*

The R'mel aquifer, located south of Larache city along the Atlantic coast, constitutes an essential groundwater tank. It varies originally and seasonally from good chemical quality to very poor quality. The R'mel aquifer is shallow in the south, where the level is roughly 5 m below the ground. The water table in the north and the littoral zone varies between 15 and 20 m. It is used to supply drinking, industrial, and irrigation water. The drinking water supply is done by individual, artesian, or surface wells. The R'mel aquifer is experiencing intensive pumping to supply drinking water to the rural inhabitants and for the irrigation of agricultural lands. The R'mel aquifer is mainly fed by precipitation and irrigation water. The phenomenon of the water table upwelling, observed at the level of the R'mel plateau, is mainly due to the over-irrigation and the lack of a suitable drainage system.

In the region of R'mel, the sandy-textured plateaus have an excessively low water-retention capacity and limited fertility. As a result, the R'mel region has become more exposed to pollution by several pollutants deriving from the expansive use of inorganic fertilizers and pesticides. Because of the excessive groundwater pumping, the problem of seawater intrusion is another growing concern in the region. Legislative and regulatory dispositions on soil and their protection within the irrigated areas are few and scattered. A large part of R'mel lands has been developed for crops under rotations and sprinkler irrigation. The aspect of crop rotation in the R'mel region has resulted in maximum use of the soil and water. Hmamou and Bounakaya [40] pointed out that water resources in the R'mel region have become insufficient to meet irrigation and other needs on agricultural land. Added to this is farmers' lack of understanding regarding the excessive spread of fertilizers and pesticides in the R'mel area.

*2.3. Sample Collection and Preparation*

Soil samples were collected from the R'mel agricultural area within the Loukkos perimeter located in the northwest of Morocco between the city of Tangier and Rabat. The samples were taken from three agricultural fields cultivated for potatoes and strawberries. The three chosen cities were recently fertilized. The sampling was carried out manually using a shovel at a level between 0 and 30 cm from the surface of the fields. For each of the three sites, a quantity in the order of 5–8 kg was taken. The samples were dried, sieved to 2 mm, and preserved immediately at a temperature below 4 °C in polyethylene plastic bags.

*2.4. Soil Sample Analysis*

Soil samples were dried at room temperature and then sieved using a vertical sieving machine to separate and sort grain-size fractions. The pH was measured following METHOD 9045D. Loss on ignition (LOI) analysis was used to determine the organic matter content (%OM) of the three soil samples. The colorimetric molybdenum blue method was used to determine the available phosphorus in soil samples as orthophosphate after digestion by $HNO_3$-HCL 1:3 ($v/v$). The cation exchange capacity (CEC) is related to the soil's clay and organic matter content. This measurement makes it possible to know the total quantity of exchangeable cations ($K^+$, $Ca^{2+}$, $Mg^{2+}$, $Na^+$, $H^+$, etc.) tending to retain the nutrients and phytosanitary products available to plants. The CEC of the soils was determined using the Metson method [41], which is based on the extraction of cations by 1N ammonium acetate at pH 7.0.

Major elements such as Ca, Fe, K, Mg, and Na, and heavy metals such as, Cd, Co, Cr, Cu, Mn, Mo, Ni, Pb, Se, Ti, and Zn were determined in the fine fraction of the soil (<63 μm fraction) using inductively coupled plasma–optical emission spectroscopy (ICP-OES, Agilent 5100, Tokyo, Japan). Analysis was performed after acid digestion ($HNO_3$-HCL 1:3 ($v/v$)) using a DigiPREP blocks digestion and heating system (SCP Science, Montreal, QC, Canada). The concentrations of heavy metals were expressed on a dry mass basis (mg/kg). Given that the sampled soils were characterized by a coarse texture mainly dominated by sand particles, trace elements and macronutrients were analyzed on the fraction below 63 μm. From another point of view, the objective of analyzing the fine fraction of the soil despite its low percentage is to give an idea about the role of the silty and clayey fraction in the retention of heavy metals in sandy soils.

## 2.5. Olive Pomace Biomass Slag (OPBS)

During biomass combustion, two types of waste are generated: bottom ash or slag and fly ash. Biomass slag comprises the coarser fraction of ash produced on the grate in the primary combustion chamber. The residual ash forms molten aggregates that are not transported out of the burner grate and/or furnace, thus forming slag [42]. The presence of alkali metals in biomass decreases the melting point of ash, allowing for faster slag formation [43]. The biomass slag is often mixed with mineral impurities contained in biomass fuel, such as sand, stones, and mud, or with bedding material in fluidized bed combustion plants. These impurities can be mineral, especially in fixed-bed combustion plants, and give rise to slag formation (due to a lowering of the melting point) and the presence of sintered ash particles in the bottom ash. In this study, biomass slag residue was sampled from a biomass-fired power plant for the combustion of olive pomace (Figure 1). The raw biomass slag residue was air dried at room temperature, crashed manually, and then sieved at 2 mm mesh size. The pH of the OPBS was measured following the 9045D method using a Thermo Orion pH meter (Waltham, MA USA) equipped with a low-sodium-error electrode. For trace-element determination in the OPBS, the sample was digested by $HNO_3$-HCL 1:3 ($v/v$) using a DigiPREP blocks digestion and heating system. The leached trace elements were analyzed by direct injection in (ICP/OES). The moisture content or humidity percentage in the OPBS was calculated by measuring the loss in weight after drying the sample at 105 °C for 24 h. The unburned carbon fraction in the collected Biomass Slag was quantified by the loss on ignition (LOI) method.

## 2.6. Soil and Olive Pomace Biomass Slag (OPBS) Characterization

Various characterization techniques were used in this study in order to compare the physical and chemical properties of soil samples and OPBS. The Brunauer–Emmett–Teller (BET) analysis was used to determine the adsorption characteristics such as N adsorption–desorption curves, specific surface area, and porosity of the soils and (OPBS) under N2 adsorption at 77 K using the Micromeritics Tristar II 3020 Surface Area Analyzer (Micromeritics Instr. Corps., Norcross, GA, USA). Scanning electron microscopy (SEM) was employed to determine the morphology of the soil samples and OPBS using a (SEM, Hitachi, Tokyo, Japan, S-4800). At the same time, an x-ray fluorescence spectrometer (XRF, PANalytical Axios FAST simultaneous WDXRF, Malvern PANalytical Ltd., Almelo, The Netherlands) was used to determine the mineral composition of the fine-soil fraction (<63 μm fraction) and OPBS. The crystalline phases of the soil's complete fraction and OPBS were determined by X-ray diffraction (XRD) using an X-ray diffractometer PANalytical

X'Pert Pro (Malvern PANalytical Ltd., Almelo, The Netherlands). Thermo-gravimetric and differential calorimetric scanning TGA/DCS analysis of the soil's fine and complete fractions was carried out using an SDT Q600 V20.9 Build 20 (TA Instruments, Newcastle, DE, USA).

**Figure 1.** Slag residue from olive pomace combustion (OPBS).

*2.7. Column Study*

In order to assess the soil retention capacity of $NO_3^-$ in the R'mel area, a study of nitrate leaching via vertical columns was carried out. A total of six columns were used in this study. The glass columns used for this study had a length of 35 cm and a diameter of 5 cm. The soil profiles were brought to the field conditions by controlling the apparent density of the soil by adding the necessary amounts of water to promote the vertical movement of solute. Generally, sandy soils are characterized by their high permeability allowing the effluent to infiltrate by gravity and do not necessarily need a force to be drained. All the experiments consisted of the addition of 100 mL constant daily volume of $N-NO_3^-$. In the first experiment, soils 1, 2, and 3 were filled in columns, namely C1, C2, and C3, conditioned with ultrapure water ($T_0$), and an initial concentration of 51.60 mg/L $N-NO_3^-$ was added on a daily basis for 17 days. In the second experiment, the columns labeled C4, C5, and C6 were filled with soils 1, 2, and 3, respectively. Before initiating the experiment, the three columns were flushed several times with ultrapure water in order to remove the excess nitrogen already contained in the samples. Afterward, the three columns were loaded with an increasing concentration of $N-NO_3^-$ ranging from 0 to 102.83 mg/L. Figure 2 shows the initially loaded concentrations for experiments 1 and 2. The collected solutions were analyzed every day for TN and expressed as $N-NO_3^-$. Figure 3 shows the column setup for the nitrate-leaching experiments of the R'mel soils.

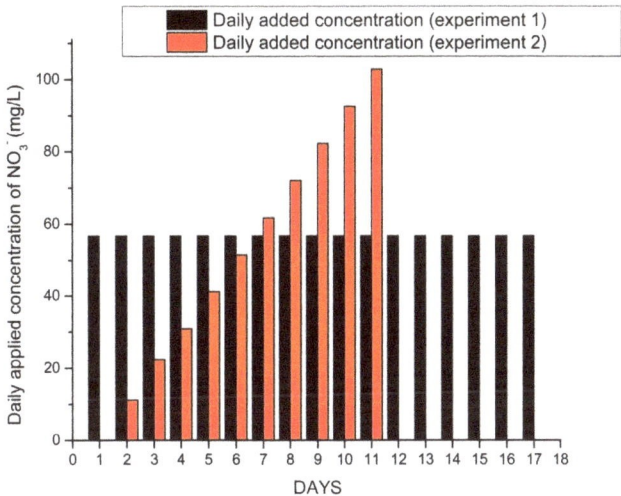

**Figure 2.** Daily applied concentration of nitrate during experiment 1 (C1, C2, and C3) and experiment 2 (C4, C5, and C6).

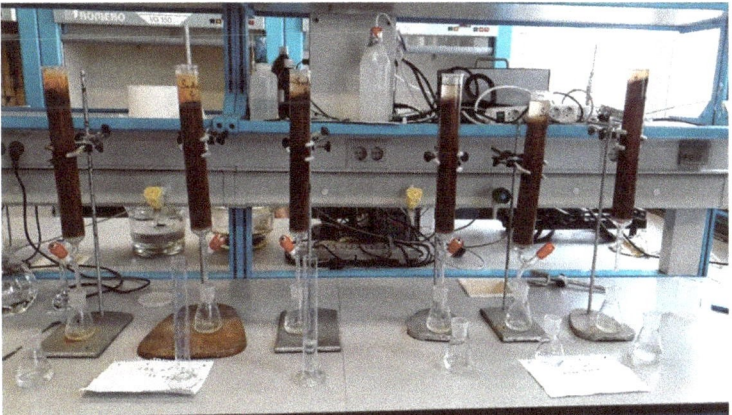

**Figure 3.** Column setup for the nitrate-leaching experiments of the R'mel soils.

*2.8. Measurements and Data Analysis*

The total nitrogen (TN) measurements were conducted by catalytic combustion at 720 °C using a Shimadzu TOC–VCSH analyzer, according to the manufacturer's instructions, with a 5.0% coefficient of variation. A total of six columns were prepared, and 82 samples were measured for TN. $NH_4^-$ and $NO_2^-$ were considered negligible in this study because of their weak concentrations in the R'mel groundwater [39]. The higher $N-NO_3^-$ concentrations reported in the study area were mainly due to the shallow depth of the water table and the aerated conditions allowing the continued oxygenation of the soil and groundwater. According to Zarabi and Jalali [44], $N-NO_3^-$ is the predominant form in N leachate solution due to its high solubility and lower affinity to be adsorbed by soil sites. The following conversion equation was used to express nitrate concentration in the leached solution.

$$\text{Nitrate} - NO_3 \left(\frac{mg}{L}\right) = 4.4268 \times \text{Nitrate} - N \left(\frac{mg}{L}\right)$$

## 3. Results
### 3.1. Soil Physicochemical Characterization

The results of the physicochemical properties of the soil are shown in Table 1. The granulometric classification made it possible to classify the three types of soil according to their particle size into three groups: sand, silt, and clay. The grain-size distribution showed that the soils have a sandy texture (>95% of sand) with low silt and clay content. This granulometric classification highlights the infertility of the R'mel soils, which leads farmers to intensify the use of inorganic fertilizers and manures in order to increase soil fertility. In consequence, the intensive irrigation rates applied to crops and soils have caused the leaching of these amendments towards groundwater due to the soil's low capacity to adsorb fertilizers.

Table 1. Physicochemical analysis of soil samples.

| Parameter | S01 | S02 | S03 | Mean |
|---|---|---|---|---|
| pH | 7.27 | 8.75 | 6.33 | 7.55 |
| EC (mS/Cm) | 0.26 | 0.23 | 0.32 | 0.27 |
| OM % CF | 2.81 | 2.36 | 1.57 | 2.25 |
| OM % FF | 9.11 | 7.5 | 7.65 | 8.09 |
| CEC (meq/100 g) | 9.28 | 9.11 | 8.18 | 8.86 |
| Bulk density | 1.28 | 1.34 | 1.32 | 1.31 |
| Sand % | 95.84 | 95.87 | 95.42 | 95.71 |
| Silt % | 1.18 | 2.47 | 2.33 | 1.99 |
| Clay % | 2.98 | 2.36 | 1.66 | 2.33 |
| $PO_4^{3-}$ (g·Kg$^{-1}$) * | 1.26 | 2.04 | 0.81 | 1.37 |
| Ca (g·Kg$^{-1}$) * | 7.5 | 13.37 | 5.24 | 8.7 |
| Fe (g·Kg$^{-1}$) * | 43.76 | 47.41 | 46.19 | 45.79 |
| K (g·Kg$^{-1}$) * | 3.48 | 3.158 | 2.05 | 2.9 |
| Mg (g·Kg$^{-1}$) * | 4.58 | 4.86 | 3.42 | 4.29 |
| Mn (g·Kg$^{-1}$) * | 1.62 | 1.35 | 1.53 | 1.5 |
| Na (g·Kg$^{-1}$) * | 0.33 | 0.27 | 0.18 | 0.26 |
| As (mg·Kg$^{-1}$) * | 62.3 | 53.3 | 58 | 57.9 |
| Cd (mg·Kg$^{-1}$) * | 6.8 | 2.64 | 1.48 | 3.6 |
| Co (mg·Kg$^{-1}$) * | 22.1 | 20.3 | 32.6 | 25 |
| Cr (mg·Kg$^{-1}$) * | 123.5 | 125.4 | 89.6 | 112.8 |
| Cu (mg·Kg$^{-1}$) * | 108.1 | 38.9 | 17.1 | 54.7 |
| Mo (mg·Kg$^{-1}$) * | 2.84 | 0.82 | 0.97 | 1.5 |
| Ni (mg·Kg$^{-1}$) * | 52.4 | 36.6 | 48.6 | 45.9 |
| Pb (mg·Kg$^{-1}$) * | 37.3 | 23.1 | 27.6 | 29.4 |
| Zn (mg·Kg$^{-1}$) * | 224.7 | 162.14 | 114.25 | 167 |

* analyzed in the fine fraction (FF) of soil.

These results were also confirmed by analyzing the organic matter available in the three soils. Indeed, analyses of the organic matter in the complete fraction, CF, of the soil showed that the R'mel soils were characterized by a low OM content of 2.8, 2.36, and 1.57% for the samples S01, S02, and S03, respectively. On the contrary, the fine fraction, FF, which represents the fraction under 63 µm mainly composed of silt and clay, has a higher OM content. Soil 1 was characterized by the highest OM rate of 9.11%, whereas percentages of 7.5 and 7.65% were measured in S02 and S03, respectively. The pH of the R'mel soils varied between 6.33 and 8.7, with a mean value of 7.55. The soil S02 was alkaline in nature with the highest pH value of 8.75; soil S03 was slightly acidic with a pH value of 6.33, whereas soil S01 had a neutral pH value of 7.57. The measured cation exchange capacity (CEC) showed values of 9.28, 9.11, and 8.18 meq/100 g for the samples S01, S02, and S03 with a mean value of 8.86 meq/100 g of soil. This value confirms the low percentage of soil organic matter and clay in the R'mel soil. The electrical conductivity, EC, of the soil indicated the salinity of the sampled soils and ranged between 0.23 and 0.32 mS/Cm, thus indicating a low EC. Phosphorus in soils is almost entirely in the form of orthophosphate, with total P concentrations typically ranging from 500 to 800 mg/kg dry soil [45]. The presence of orthophosphate in the soil is closely related to organic matter and clay minerals. The analysis of $PO^{3-}_4$ in the soil's fine fraction showed a concentration ranging from 0.81 to 2.04 g/Kg, which indicates that despite the dominance of sandy texture, the clayey fraction tends to adsorb higher amounts of orthophosphate ions. Nutrient availability in soils is affected by many interconnected variables; examples include parental rock composition, particle size, humus and water content, pH, aeration, temperature, root surface area, and fungal growth [46]. The lack of nutrients in sandy soils is frequently resulting in decreased water-holding capacity, soil pH, cation exchange capacity, and soil organic matter [47]. The concentrations of nutrients in the fine fraction of the R'mel soils were in the following order: Fe > Ca > Mg > K> Mn > Na, with values ranging from 46.19 to 47.41, 5.24 to 13.37, 3.42 to 4.86, 2.05 to 3,48, 1.35 to 1.62, and 0.18 to 0.33 g/Kg, respectively. The lack of OM in the R'mel soils probably influences the plants' nutrient availability since the dominant texture is coarse. The higher $Ca^{2+}$ content (13.37 g/kg) recorded in Soil 2 indicates that calcium cation influences the pH of the soil. This relatively higher alkalinity in Soil 2 could be originating from the application of free lime (CaO) to increase soil pH for optimal plant growth in this area known for its low alkalinity. The concentrations of these nutrients in the fine fraction of the R'mel soil indicate that their presence is barely at the level recommended for agricultural soils. Furthermore, this fraction accounts for less than 5% of the total soil fraction, although the remaining fraction is notably coarse particles with low CEC and low water- and organic-matter-holding capacity, and therefore poor in nutrients.

During their assimilation, certain trace elements, such as Cu, Zn, Ni, Fe, Co, Se, and Ba, are essential for the functioning of plants [48]. They intervene in processes such as photosynthesis, biosynthesis of proteins, amino and nucleic acids, and chlorophyll, as well as the production of substances made by plants that make them competitive in their environment [49]. However, soil pollution by heavy metals is one of the world's major environmental problems [50]. The accumulation of heavy metals in soil could originate from geogenic or/and anthropogenic sources. Several studies have shown that agricultural practices could be a source of heavy metal accumulation in agricultural fields. Several studies [51–53] reported that the applications of fertilizers and pesticides were responsible for the accumulation of Cr, Cd, Cu, Zn, Ni, Mn, and Pb in agricultural fields.

Table 1 represents the concentrations of trace elements analyzed in the fine fraction of the R'mel soils. The heavy metals were well absorbed by the FF of the soil, which was evident from the As, Cr, Cu, and Ni contents surpassing the allowable limits set at 20 ppm, 0.8 ppm, 30 ppm, and 35 ppm, respectively. Zn concentrations varied between 114 and 224 mg/kg in the soil samples. Zn is generally present in soils at background concentrations of 10–100 mg Zn $kg^{-1}$ [54]. Arsenic is a ubiquitous element that can be found in every Earth compartment; arsenic derives naturally from geogenic rocks and/or can originate from anthropogenic activities such as the use of fertilizers and pesticides. R'mel drinking water was already reported as being contaminated by arsenic. This contamination has been attributed to both geogenic and anthropogenic sources [39]. From these results, it is possible to mention the crucial role of the fine fraction (clays and silts) in the retention of pollutants and also to shed light on the impact of low organic matter on the release of pollutants towards groundwater. By comparing the amount of clay in the three samples, it was consistent that Soil 1 had retained a higher concentration of trace elements than Soils 2 and 3; this higher concentration is closely related to the higher amount of OM and clay contained in the soil which is manifested by the relatively higher adsorption capacity of trace elements. Despite this finding, the R'mel soils contain negligible amounts of silt and clay and consist largely of coarse particles, which affects the leaching of trace elements and other agricultural pollutants.

*3.2. Column-Leaching Experiments*

The nitrate-leaching concentrations recorded during experiment 1 are shown in Figure 4a. The first day of leaching was marked by elevated leaching rates of 96.70, 116.82, and 184.82 mg/L for C1, C2, and C3, respectively. The measured concentration indicated that leaching exceeded the initially loaded concentrations and imply that nitrogen was already present in the soil samples from field fertigation activities. It was also evident that Soil 3 had a higher nitrogen content. On the second day, there was an evident decrease in $NO_3^-$ concentration in C3 (77.52 mg/L), whereas $NO_3^-$ concentration in C1 and C2 decreased slightly to 91.25 and 94.2 mg/L, respectively. Except for day one, the daily measured $NO_3^-$ concentration in C3 was lower than in C1 and C2 until day five, when concentrations were comparable at 70.61, 68.66, and 67.55 mg/L in C1, C2, and C3, respectively. The observed leachate concentrations dropped consistently until day eight, indicating that $NO_3^-$ loading was roughly close to the leachate concentration in Soil 3 (55.73 mg/L). The $NO_3^-$ leaching in C1 and C2 was close to the initial loaded concentration from D11 of the experiment. Afterward, the $NO_3^-$ concentrations decreased slightly in the three columns and ranged between 56.08 and 49.47, 56.79 and 46.97, and 50.13 and 47.79 mg/L in C1, C2, and C3, respectively. The results of this experiment show that after fertigation of the raw soils with 51,56 mg/L of $NO_3^-$, the soils need more time to eliminate the $NO_3^-$ excess already contained in the soil. The last days of the experiment showed that the loaded concentration of $NO_3^-$ was moderately equal to the leached concentration. This experiment highlighted that after 17 days from the application of a constant $NO_3^-$ concentration, the leaching rates do not decrease in the collected leachate. On the contrary, the raw sampled soils contribute, in turn, to the enrichment of $NO_3^-$ in leachate. Nitrate-permissible level in groundwater was determined to not exceed 50 mg/L. This experiment showed that an excess of fertigation/irrigation rates in the R'mel area could cause the contamination of local groundwater by nitrate.

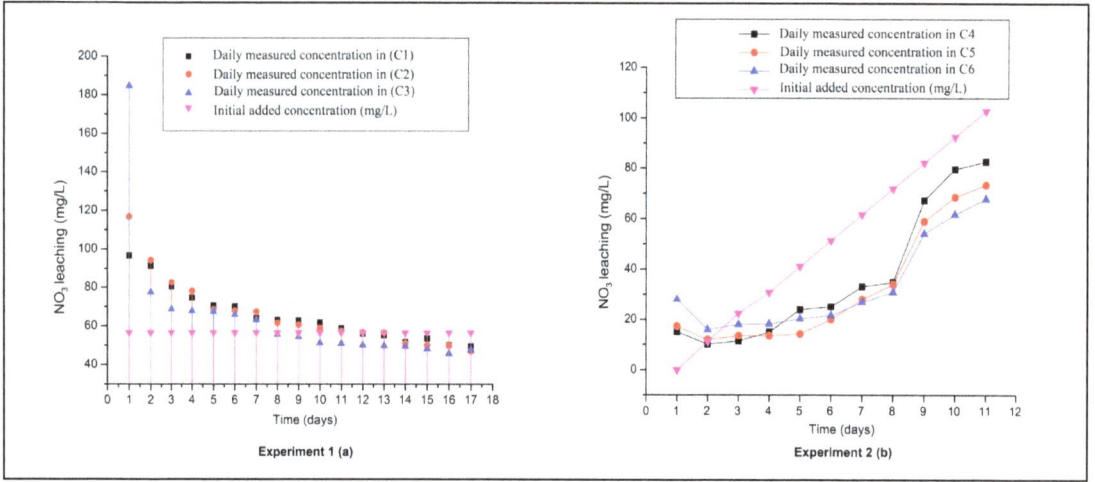

**Figure 4.** $NO_3^-$ leaching from experiment 1 (**a**) and experiment 2 (**b**).

Experiment 2, illustrated in Figure 4b, shows the effect of increased nitrate doses on leaching rates from the three sandy soils. Nitrate leaching on the first day was higher than the initially applied concentration of 0 mg/L (only water), reaching 15.1, 17.3, and 28.0 mg/L in columns C4, C5, and C6, respectively. These results indicate that although the soil columns were diluted several times with water to remove excess nitrate present in the soil, the three columns still contain nitrate, indicating residual or secondary leaching from soil samples. The following application (day two) shows that the measured NO3- concentration in C4, C5, and C6 decreased while the initially applied $NO_3^-$ increased from 0 to 11.21 mg/L. After day three, all the measured concentrations in leachates were lower than the initially applied concentration until the end of the experiment. The leaching curve increased slightly from day three to day eight for an increasing loaded concentration ranging from 30.85 to 71.98 mg/L. These results show that the maximum adsorption capacity of the soils was reached between days three and eight. Meanwhile, a rapid increase in $NO_3^-$ leaching was observed in the three columns on day nine, potentially indicating that the soils started to progressively lose their ability to adsorb nitrate.

In general, the nitrate leaching in columns C4, C5, and C6 seems to increase with respect to the applied doses of nitrate. However, experiment 2 demonstrated that after removing nitrogen excess from the raw soils, it appears that nitrate leaching decreases in the three columns. This indicates that R'mel soils have a low adsorption capacity which is quickly affected by higher nitrogen applications.

By comparing the collected leachate with the measured added concentrations in experiment 2, it is evident that, contrary to experiment 1, the $NO_3^-$ leaching did not exceed the applied $NO_3^-$ concentration. This suggests that the elevated applied N fertilizer affects the soil's adsorption capacity in the R'mel area. In this context, the following sections will be dedicated to discussing the main factors affecting the R'mel soil retention capacity through various physical and chemical characterizations.

*3.3. Soil Thermal Characterization*

The adoption of a technique such as TGA-DSC could alleviate the problem of soil decomposition and provide an accurate description of soil composition by comparing the different temperature intervals of the soil. The simultaneous (TGA-DSC) measurements for both the fine and complete fractions of the sampled soils are presented in Figure 5. The observed effects of TGA variation include three intervals; the first ranges from 0 to 105 °C and represents the loss of interstitial water from the samples intra-pores, the second interval refers to the pyrolyze/oxidation of OM under a maximum temperature of 550 °C, whereas the third interval represents the decomposition of $CaCO_3$ at a temperature exceeding 550 °C. Soil 2 was characterized by higher derivative peaks (600 °C and 700 °C) in its FF and CF; these results confirm the presence of $CaCO_3$ in this soil. This will be also confirmed by both XRD and XRF results (CaO and $CaCO_3$), in addition to the soil's higher calcium content and pH. It is noteworthy that the higher mass losses are accompanied by the higher peak of derivative mass loss, thus highlighting the significant changes in weight (inflection points) and demonstrating the areas corresponding to the decomposition of the soil during the heating process. Moreover, the points corresponding to changes in heat flow were highlighted by higher derivative peaks. The TGA/DSC of the soil FF displays a clear peak when compared with the CF. The mass loss of the FF was clearly higher than the CF which is mainly due to higher water content, organic matter, and carbonates in the clayey fraction. This difference between the TGA/DSC curves provided information on the parameters influencing the soil retention capacity, such as the presence of fine particles, water-holding capacity, and porosity.

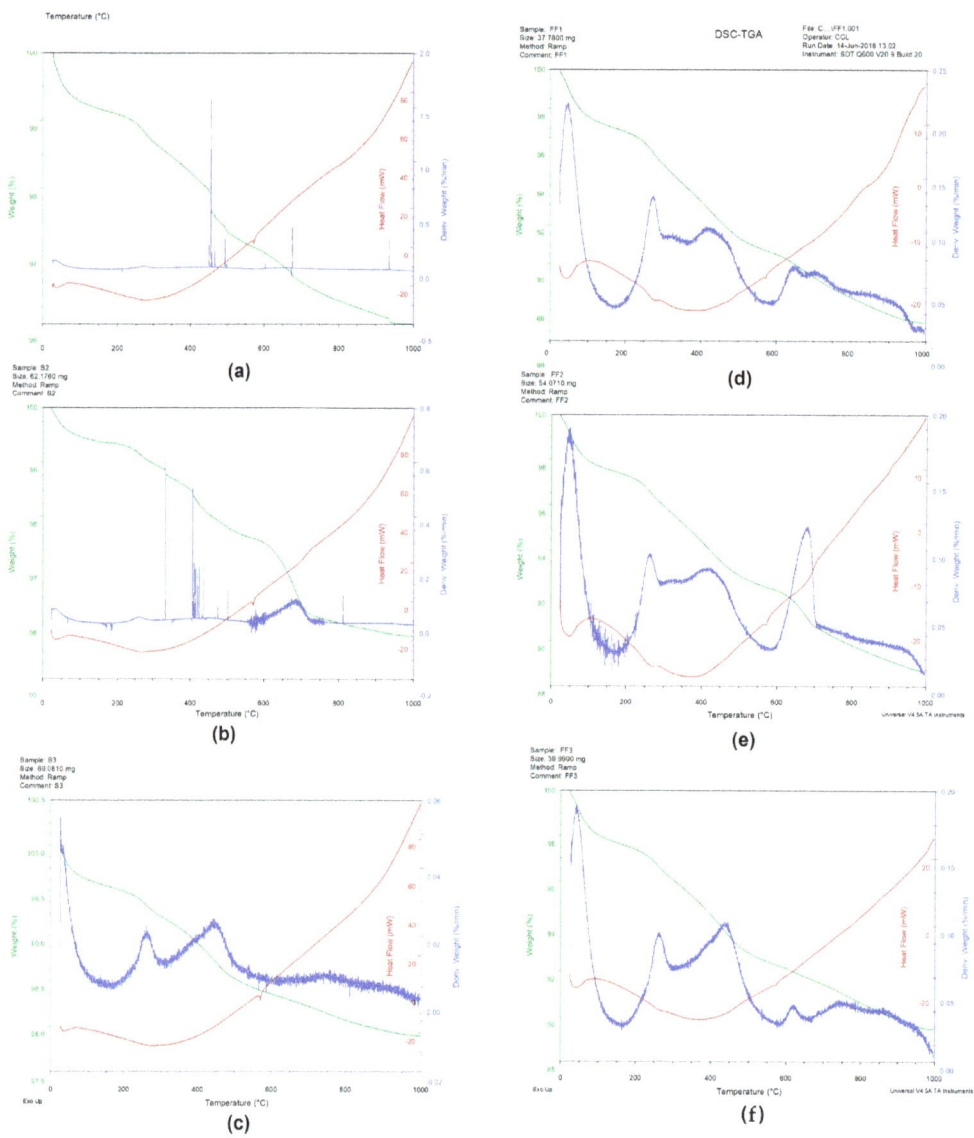

**Figure 5.** Thermal decomposition profiles of the R'mel soils: (**a**–**c**) represent the TGA/DSC curves for the complete fraction of Soils 1, 2, and 3, respectively. Images (**d**–**f**) represent the TGA/DSC curves for the fine fraction (<63 µm) of Soils 1, 2, and 3 respectively.

*3.4. Olive Pomace Biomass Slag (OPBS) Analysis*

The composition of OPBS was characterized as major (Table 2). The OPBS has a very alkaline pH of 12.1. This higher pH is related to the presence of dissolved metals as basic metal salts, oxides, and carbonates formed during the combustion of biomass. Accordingly, OPBS could be used to increase the pH of acidic soils in the R'mel region. The OPBS represents a moisture content of 7.18% calculated as dry mass. This moisture content was much higher when compared to soil samples which did not exceed 1% of the total mass based on the generated TGA curves. These findings demonstrate that the pore spaces between OPBS particles could hold more interstitial water than the R'mel soils, implying that the application of OPBS to soil may boost soil water-retention capacity. During biomass combustion, the organic carbon present in the slag corresponds to the unburned fraction of the biomass. The OPBS contains 19.97% of total organic carbon (TOC) that has not been burnt. Batra et al. [55] investigated the presence of unburned carbon in bagasse fly ash sampled from bagasse cogeneration power plants in India and found more than 25% of unburned carbon. The same authors reported that unburned carbon resulted in disposal issues, provided challenges when employed in cement formulations, and would thus be better suited for alternative uses. In this context, there is a distinct possibility to apply the OPBS as a carbon-containing amendment to R'mel sandy soils recognized by their lower organic carbon as demonstrated by thermogravimetric analysis. In addition to the high carbon content, the major element compositions of OPBS showed significant amounts of macronutrients which decrease in the sequence of Ca > K > Mg > Fe > Na. The leaching of essential macronutrients is common in coarse-textured soils and reduces nutrient availability to plants. By comparing the macronutrient level in OPBS with soil, it is convincible to exploit the fertilizing capacity of this material to amend R'mel soils. In addition, many farmers in the R'mel region use lime in order to control soil acidity mainly resulting from irrigation and excessive nitrogen use. The cationic exchange capacity (CEC) plays an important role in adsorbing and releasing nutrients needed by plants, as well as assessing the potential harm of certain contaminants. As a result, the R'mel soils have shown quite low CEC. The higher level of macronutrients contained in OPBS such as calcium, potassium, and magnesium could be introduced to increase the soil CEC, and thus macronutrient availability to the plants. OPBS, on the other hand, has the potential to limit the excessive use of N fertilizers by not only giving essential nutrients but also preserving their availability in the soil. Motesharezadeh and collaborators [56] concluded that the use of lime in sandy soils increases the CEC and reduces the leaching of $NO_3^-$ and K.

Table 2. pH, moisture content, unburned carbon, and major elements in OPBS.

| Parameter | pH | Moisture % | Unburned Carbon % | Ca % | Fe % | K% | Mg% | Na% |
|---|---|---|---|---|---|---|---|---|
| OPBS | 12.1 | 7.18 | 19.97 | 10.59 | 0.95 | 8.24 | 1.56 | 0.15 |

The combustion of biomass could result in the accumulation of trace elements in ashes and slag residues. For example, Wang et al. 2014 [57] reported that woody biomass blended in the fuels could generate large amounts of As, Cd, Cu, Cr, Pb, and Zn in fly ash and slag. In this study, OPBS was analyzed to assess trace-element toxicity in order to insure a safe utilization of this residue in soil (Table 3). The heavy metal content in OPBS decreases in the order of Mn > Zn > Cr > Cu > Ni > Co > Mo > As > Pb > Cd. It can be seen that the levels of heavy metals such as, Cd, Co, Pb, and Mo were quite low in the OPBS samples, whereas Cu, Ni, Cr, Mn, and Zn occurred in background levels, and no element present any potential risk of contamination. Moreover, the toxicity of the slags is not intrinsically linked to their trace-element levels, but rather to their leaching. A previous similar study of four biomass slags from a fired power plant showing approximately the same composition of OPBS has demonstrated too low leaching amounts of trace elements [34]. The same authors have concluded that the biomass slags did not represent any risk of contamination related to their utilization. In another study, the fertilizer value of fly ash derived from burning bark and wood chip was investigated by Numesniemi et al. [58], and the findings revealed that the levels of hazardous elements (As, Cd, Cr, Cu, Hg, Pb, Ni, and Zn) were low. In addition, the comparison between the level of trace elements in the clayey fraction of the soil and OPBS revealed that the fine fraction contained higher trace elements than the OPBS; these results indicate that the use of OPBS will not affect the levels of trace elements in the soil. From another point of view, the OPBS, which has a similar composition to fly ashes, could represent a great solution for the immobilization of heavy metals in the soil.

Indeed, several studies reported the efficiency of fly-ash addition to soil on trace-element immobilization [57–65].

Table 3. Trace element composition in OPBS.

| Trace Element in (mg. Kg$^{-1}$) | As | Cd | Co | Cr | Cu | Mn | Mo | Ni | Pb | Zn |
|---|---|---|---|---|---|---|---|---|---|---|
| OPBS | 2.12 | 0.25 | 2.49 | 48.43 | 48.30 | 396.33 | 1.72 | 42.64 | 1.93 | 47.34 |

*3.5. X-ray Fluorescence Analysis*

The major oxides present in the soil's FF, CF, and OPBS are presented in Table 4. The percentage of oxides decreased in the sequence of $SiO_2 > Al_2O_3 > Fe_2O_3 > CaO > MgO > K_2O > TiO_2 > Na_2O > P_2O_5 > SO_3$ in the three analyzed soils for both the fine fraction (FF) and the complete fraction (CF) of the soils. The results showed that the samples were predominately silicate soils rich in iron and aluminum. According to Chong et al. [66], the high content of exchangeable Al and Fe ions is mainly due to the high temperatures and heavy rains resulting in low soil pH, low nutrient availability, and low organic matter content. A significant difference in $SiO_2$ percentage was noted between the FF and CF, confirming quartz particles' predominance in the R'mel soils. In contrast, the $SiO_2$ percentage has decreased while other oxides have increased parallelly in the soil clayey fraction. The mineral matrix present in the soil ranges between 83.3% and 89.62% for the FF, whereas it ranges from 89.04 to 92.13% for the CF, indicating the higher rate of soil cultivation in the R'mel area. Considering the heterogeneity of the situation in soils, normally the denser and more friable/loose mineral fraction (less oxide/silica content) is found in the FF of the soil representing higher OM, CEC, and specific surface area (SSA). On the contrary, the less friable (or gangue) material is typically composed of oxides, often the majority being silica. In addition, the higher percentage of mineral matrix signifies a more dominant effect of hydrogen bonding on the adhesion of inorganic oxide particles. The size distribution and mineralogy of the clayey and silty fractions associated with sand grains are also responsible for variations in the physical properties of tropical sandy soils [67]. The XRF analysis of OPBS reveals a distinct composition dominated mainly by $SiO_2$, CaO, and $K_2O$ accounting for more than 66% of the total OPBS. Oxides such as CaO, $K_2O$, MgO, $P_2O_5$, and SO3 were present at negligible amounts in both soil fractions. In contrast, these oxides were found in high concentrations in the OPBS. Consequently, the OPBS can provide essential nutrients needed by crops when amended with OPBS. In addition, the higher alkali oxides represent a great advantage due to the liming characteristics of this material.

Table 4. Chemical composition of the three soil samples and Olive Pomace Biomass Slag (OPBS).

| Sample | SiO$_2$ | Al$_2$O$_3$ | Fe$_2$O$_3$ | MnO | MgO | CaO | Na$_2$O | K$_2$O | TiO$_2$ | P$_2$O$_5$ | SO$_3$ | %Mineral Fraction |
|---|---|---|---|---|---|---|---|---|---|---|---|---|
| F.F 1 | 55.17 | 10.86 | 9.9 | 0.3 | 1.57 | 1.73 | 0.78 | 1.42 | 0.75 | 0.61 | 0.21 | 83.3 |
| C.F 1 | 71.43 | 7.19 | 5.94 | 0.16 | 0.55 | 1.03 | 0.76 | 0.94 | 0.55 | 0.34 | 0.15 | 89.04 |
| F.F 2 | 58.79 | 11.09 | 9.76 | 0.23 | 1.59 | 2.56 | 0.76 | 1.38 | 0.71 | 0.53 | 0.17 | 87.57 |
| C.F 2 | 72.45 | 7.45 | 5.00 | 0.16 | 0.53 | 1.21 | 0.72 | 1.11 | 0.45 | 0.38 | 0.14 | 89.6 |
| F.F 3 | 60.64 | 12.51 | 10.74 | 0.3 | 1.3 | 1.01 | 0.51 | 1.3 | 0.81 | 0.37 | 0.13 | 89.62 |
| C.F 3 | 75.70 | 7.23 | 5.52 | 0.19 | 0.42 | 0.54 | 0.63 | 1.01 | 0.48 | 0.32 | 0.09 | 92.13 |
| OPBS | 26.29 | 3.26 | 1.07 | 0.05 | 3.52 | 22.38 | 0.56 | 17.63 | 0.12 | 2.75 | 1.41 | 79.04 |

(Note: F.F = Fine-soil fraction (<63 μm); C.F = complete-soil fraction; OPBS = Olive Pomace Biomass Slag).

*3.6. XRD Analysis*

The use of X-ray diffraction is a necessary tool for determining the different crystalline phases contained in a soil sample. Figure 6 displays the XRD pattern of the three soils (S1, S2, and S3) and Olive Pomace Biomass Slag (OPBS) samples. The soils were dominated by silicon oxide (SiO2), with no other crystalline phase detected. Except for Soil 02, which displayed a minor peak of $CaCO_3$ as validated by TGA curves and XRF, which had a high proportion of CaO oxide (2,56%) when compared to Soils 1 and 3. The strongest quartz peaks were found between 20.71° and 26.45° for both the soil and the OPBS samples. XRD analysis of OPBS, on the other hand, revealed that the most prevalent crystalline phase was calcite ($CaCO_3$) and quartz ($SiO_2$), along with other minerals such as dolomite ($CaMg, (CO_3)_2$), fairchildite ($K_2Ca (CO_3)_2$), and free lime (CaO). The results of XRD confirm that higher percentages of $SiO_2$, CaO, and $K_2O$ detected by XRF analysis were the main elemental composition of the crystalline phases of OPBS. In sandy soils, significant weathering occurs at depth resulting in mineralogy where quartz is the dominant mineral in the sand and silt fraction and forms a considerable proportion of the clay fraction. It is noted that despite the prevalence of silica in the soil samples, other minerals were also detected by XRF in FF at a considerable percentage, e.g., $Al_2O$ and $Fe_2O_3$; however, the XRD patterns revealed only the quartz ($SiO_2$) which is mainly due to the absence of clayey minerals such montmorillonite, bentonite, kaolinite, etc. [67]. As a result, the application of OPBS as a soil additive might influence the weathering of Si. Matichenkov et al. [68] reported that soil properties such as P, Al, heavy-metal behavior, and adsorption capacity are governed by soil silicon compounds.

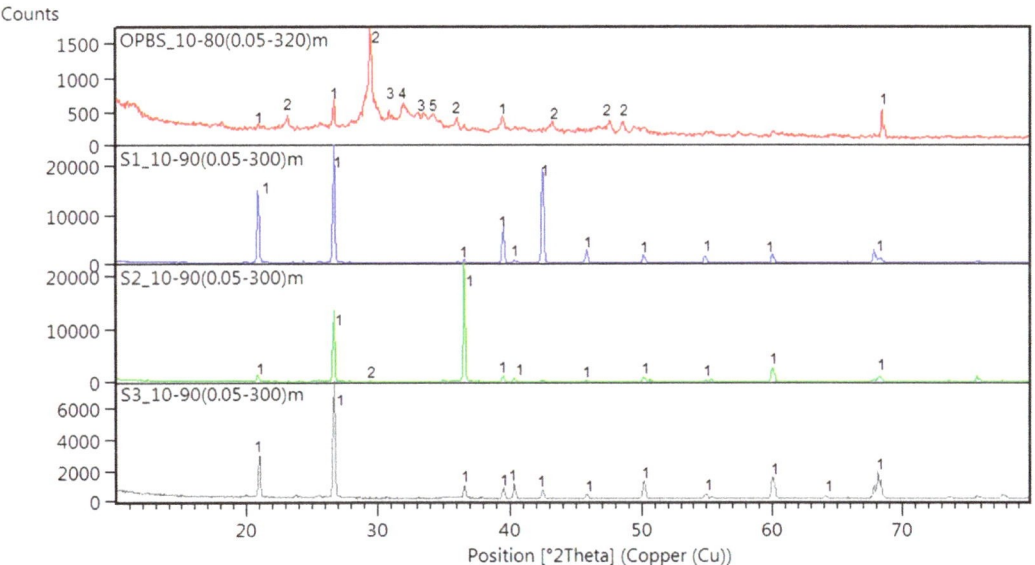

**Figure 6.** XRD analysis of the three soils, S1, S2, and S3, and OPBS: 1 = ($SiO_2$), 2 = ($CaCO_3$), 3 = ($CaMg, (CO3)_2$) 4 = ($K_2Ca (CO_3)_2$), 5 = (CaO).

*3.7. BET Characterization of the Soil Samples and Biomass Slag*

In order to investigate the surface characteristics of soil and OPBS, the BET surface area measurements were performed to provide information such as adsorption isotherms, specific surface area, pore volumes, and pore diameter. The collected gas adsorption/desorption isotherms of the samples are shown in Figure 7. The shape of isotherms corresponds to type IV which represents a mesoporous surface in which capillary condensation occurs. A hysteresis is generally observed between the adsorption and desorption curves. According to the gas adsorption isotherms, the OPBS has an adsorption volume of 33.05 ($cm^3$/g STP) at the relative pressure ($P/P_0 = 0.99$), whereas soil samples have maximum adsorption capacities of 7.70, 7.34, and 6.80 ($cm^3$/g STP) for S1, S2, and S3, respectively. This result demonstrates the high adsorption capacity of OPBS when compared with soil samples. The specific surface area represents the total area divided by the mass unit (g) and refers to the gas adsorption rate into the available pores in low-temperature conditions. The average pore diameter shows that OPBS has a higher pore diameter of 20.73 (nm) compared to soil samples. It is common for particles with smaller pores to have a higher specific surface area, but the surface area of a given particle is also determined by the number of pores in that particle, i.e., its porosity. Consequently, a particle can have very small pores, but only in a limited number, resulting in a small specific surface area as demonstrated for Soils 1, 2, and 3 (Table 5). All the samples were dominated by mesopores as shown in the pore diameter distribution (Figure 7). However, the OPBS adsorption in the function of pore volume was greater than in soil samples indicating the higher number and volume of mesopores in OPBS compared to soil samples.

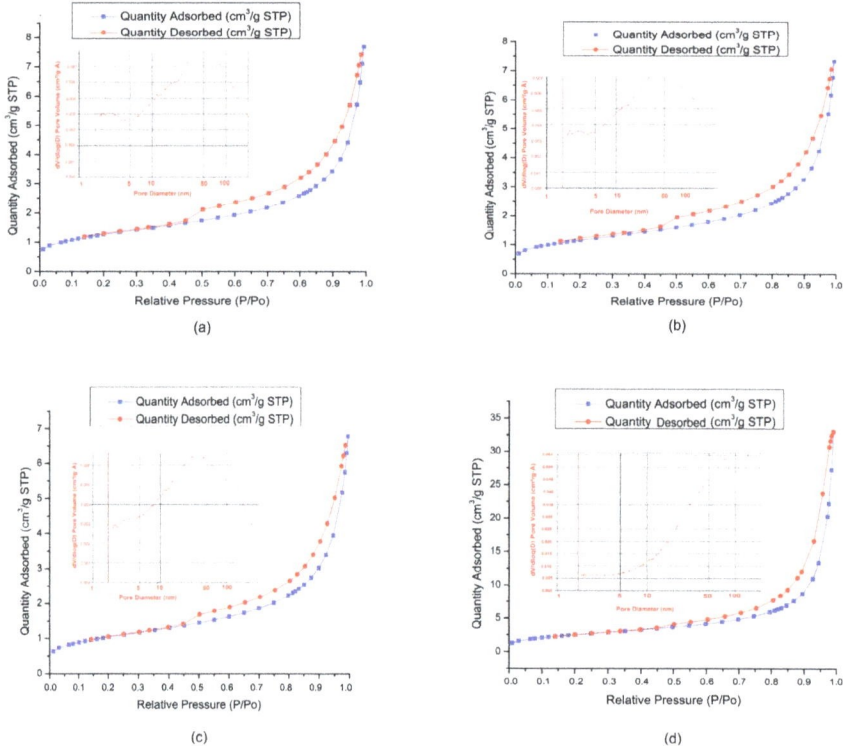

**Figure 7.** N$_2$ adsorption–desorption isotherms and pore size distribution diagrams of (**a**) Soil 1, (**b**) Soil 2, (**c**) Soil 3, and (**d**) OPBS.

**Table 5.** BET specific surface area (SSA) and pore volumes of soil samples.

| Sample | S1 | S2 | S3 | Biomass Slag |
|---|---|---|---|---|
| BET surface area (m$^2$/g) | 4.56 | 4.18 | 3.74 | 9.37 |
| Langmuir surface area (m$^2$/g) | 6.30 | 5.79 | 5.20 | 13.20 |
| Total pore volume (cm$^3$/g) | 0.008868 | 0.008574 | 0.008034 | 0.031480 |
| Pore diameter (nm) | 11.9869 | 12.3592 | 12.5881 | 20.73 |

*3.8. SEM Characterization*

In order to explore the surface characteristics (texture, pore, and pore size) of soils and OPBS, SEM analyses are presented in Figure 8. Micrographs (d, e, and f) indicate that the soil samples had approximately the same size with a prismatic shape. The micrograph scale indicates that soil particles were large in diameter and display low porosity surfaces (micrographs d and e). Indeed, the absence of laminated and porous structures indicates the presence of clayey minerals and confirms the dominance of quartz in the soil samples.

However, tiny layers have covered the quartz mineral in minor portions in some spots with a relatively low surface porosity (micrograph f). Unlike soil samples, the OPBS SEM observations demonstrate a larger porous area (micrographs a and b). In addition, the OPBS surface was characterized by a hollow surface with different pore size distributions (micrograph c). The SEM micrograph confirmed that the R'mel agricultural soils had principally a macrostructure composed mostly of larger $SiO_2$ particles, as well as a reduced pore volume (macropores). In contrast, SEM analysis revealed the abundance of pore distribution on the OPBS surface. Finally, the OPBS morphological characteristics could be useful to fill the voids between soil particles and thereby increase the surface area and water-retention capacity of soil. Moreover, the available mesopore sites on the OPBS surface could interact with the soil–water solution and adsorb different pollutants.

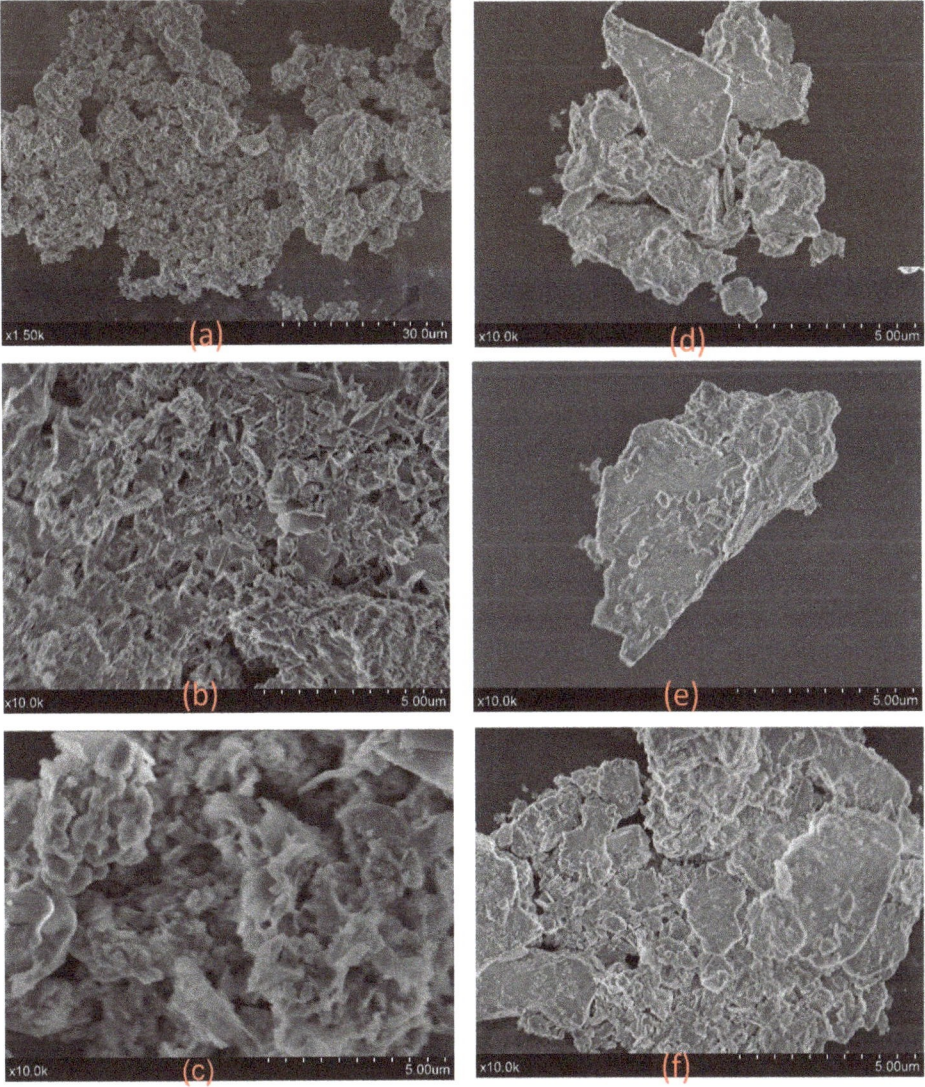

**Figure 8.** SEM micrographs (different magnifications) of OPBS (**a–c**) and soils (**d–f**).

## 4. Conclusions and Perspectives

The properties of three agricultural soils and olive pomace biomass slag (OPBS) were examined in this study to explore the possible use of OPBS as a soil additive and improver. The important findings are summarized in the following points:

(1) The R'mel soils were coarser in texture, with low clay, silt, and OM, low CEC, limited adsorption sites, and poor nutrient availability. In contrast, the clayey fraction (FF) exhibited significant water content, OM, $CaCO_3$, and heavy-metal adsorption capacity despite its low percentage in the soil (<5%).

(2) The column experiments demonstrated that the R'mel soils had a low water- and $NO_3^-$ retention capacity. Higher leaching rates in percolates were measured, even above the loaded quantities in experiment 1.

(3) The examination of OPBS showed that this residue is non-toxic, has a significant amount of essential plant nutrients such as potassium and calcium, has a moderately porous internal structure, includes organic carbon, and has a high water-retention capacity.

(4) The spreading of OPBS in R'mel soils might be supported by its higher agronomic value as a source of fertilizing elements necessary for plants (Ca, P, K, and C). OPBS can also have a direct/indirect effect on soil properties by improving the physical and chemical characteristics such as water-holding capacity, CEC, and adsorption capacity, and could contribute to the immobilization of trace elements in the soils.

Monitoring the soil and water quality in the R'mel area is currently not subject to any regulatory control. Moreover, very little data exists on the quality of groundwater and soil. The nature of the soil and the extent of the water table at a shallow depth make these waters sensitive to pollution by the surrounding agricultural activities. Therefore, it is certain that if nothing changes in the current behavior of farmers in the short term, the R'mel zone will be exposed to significant serious effects. This study intended to explain the main factors affecting groundwater quality in this area as well as propose a low-cost solution allowing the remediation and optimization of soil quality.

**Author Contributions:** Conceptualization, O.S. and F.E.M.; methodology, O.S.; software, O.S.; validation, O.S., F.E.M. and M.S.; formal analysis, J.B.; investigation, O.S.; resources, F.E.M., E.O. and J.M.; writing—original draft preparation, O.S. and F.E.M.; writing—review and editing, E.O., A.O., J.M. and F.E.M.; visualization, J.B.; supervision, M.S. and A.O.; project administration, M.S. and J.B.; All authors have read and agreed to the published version of the manuscript.

**Funding:** This research received no external funding.

**Institutional Review Board Statement:** Not applicable.

**Informed Consent Statement:** Not applicable.

**Data Availability Statement:** Not applicable.

**Acknowledgments:** The authors are thankful to Shimadzu and Merck Life Science Corporations for their continuous support.

**Conflicts of Interest:** The authors declare no conflict of interest.

## References

1. Neina, D. The Role of Soil PH in Plant Nutrition and Soil Remediation. *Appl. Environ. Soil Sci.* **2019**, *2019*, 5794869. [CrossRef]
2. Jones, A. *Soil Atlas of Africa*; Publications Office of the European Union: Luxembourg, 2013.
3. Keesstra, S.D.; Geissen, V.; Mosse, K.; Piiranen, S.; Scudiero, E.; Leistra, M.; van Schaik, L. Soil as a Filter for Groundwater Quality. *Curr. Opin. Environ. Sustain.* **2012**, *4*, 507–516. [CrossRef]
4. Panagos, P.; Borrelli, P.; Poesen, J.; Ballabio, C.; Lugato, E.; Meusburger, K.; Montanarella, L.; Alewell, C. The New Assessment of Soil Loss by Water Erosion in Europe. *Environ. Sci. Policy* **2015**, *54*, 438–447. [CrossRef]
5. Abd-Elmabod, S.K.; Bakr, N.; Muñoz-Rojas, M.; Pereira, P.; Zhang, Z.; Cerdà, A.; Jordán, A.; Mansour, H.; De la Rosa, D.; Jones, L. Assessment of Soil Suitability for Improvement of Soil Factors and Agricultural Management. *Sustainability* **2019**, *11*, 1588. [CrossRef]

6. Amer, R. Spatial Relationship between Irrigation Water Salinity, Waterlogging, and Cropland Degradation in the Arid and Semi-Arid Environments. *Remote Sens.* **2021**, *13*, 1047. [CrossRef]
7. Bhattacharyya, R.; Ghosh, B.N.; Mishra, P.K.; Mandal, B.; Rao, C.S.; Sarkar, D.; Das, K.; Anil, K.S.; Lalitha, M.; Hati, K.M. Soil Degradation in India: Challenges and Potential Solutions. *Sustainability* **2015**, *7*, 3528–3570. [CrossRef]
8. Taghizadeh-Toosi, A.; Hansen, E.M.; Olesen, J.E.; Baral, K.R.; Petersen, S.O. Interactive Effects of Straw Management, Tillage, and a Cover Crop on Nitrous Oxide Emissions and Nitrate Leaching from a Sandy Loam Soil. *Sci. Total Environ.* **2022**, *828*, 154316. [CrossRef]
9. Cui, X.; Wang, J.; Wang, J.; Li, Y.; Lou, Y.; Zhuge, Y.; Dong, Y. Soil Available Nitrogen and Yield Effect under Different Combinations of Urease/Nitrate Inhibitor in Wheat/Maize Rotation System. *Agronomy* **2022**, *12*, 1888. [CrossRef]
10. Padilla, F.M.; Gallardo, M.; Manzano-Agugliaro, F. Global Trends in Nitrate Leaching Research in the 1960–2017 Period. *Sci. Total Environ.* **2018**, *643*, 400–413. [CrossRef]
11. Lu, B.; Liu, X.; Dong, P.; Tick, G.R.; Zheng, C.; Zhang, Y.; Mahmood-UI-Hassan, M.; Bai, H.; Lamy, E. Quantifying Fate and Transport of Nitrate in Saturated Soil Systems Using Fractional Derivative Model. *Appl. Math. Model.* **2020**, *81*, 279–295. [CrossRef]
12. Craswell, E. Fertilizers and Nitrate Pollution of Surface and Ground Water: An Increasingly Pervasive Global Problem. *SN Appl. Sci.* **2021**, *3*, 518.
13. Stanley, J.; Reading, L. Nitrate Dynamics in Groundwater under Sugarcane in a Wet-Tropics Catchment. *Heliyon* **2020**, *6*, e05507. [CrossRef] [PubMed]
14. Luo, X.; Kou, C.; Wang, Q. Optimal Fertilizer Application Reduced Nitrogen Leaching and Maintained High Yield in Wheat-Maize Cropping System in North China. *Plants* **2022**, *11*, 1963. [CrossRef]
15. Ingraham, P.A.; Salas, W.A. Assessing Nitrous Oxide and Nitrate Leaching Mitigation Potential in US Corn Crop Systems Using the DNDC Model. *Agric. Syst.* **2019**, *175*, 79–87. [CrossRef]
16. Wongsanit, J.; Teartisup, P.; Kerdsueb, P.; Tharnpoophasiam, P.; Worakhunpiset, S. Contamination of Nitrate in Groundwater and Its Potential Human Health: A Case Study of Lower Mae Klong River Basin, Thailand. *Environ. Sci. Pollut. Res.* **2015**, *22*, 11504–11512. [CrossRef]
17. Huang, J.; Hartemink, A.E. Soil and Environmental Issues in Sandy Soils. *Earth-Sci. Rev.* **2020**, *208*, 103295. [CrossRef]
18. Karathanasis, A.D.; Harris, W.G. Quantitative Thermal Analysis of Soil Materials. In *Quantitative Methods in Soil Mineralogy*; Soil Science Society of America: Madison, WI, USA, 1994; pp. 360–411.
19. Li, C.; Zhou, K.; Qin, W.; Tian, C.; Qi, M.; Yan, X.; Han, W. A Review on Heavy Metals Contamination in Soil: Effects, Sources, and Remediation Techniques. *Soil Sediment Contam. Int. J.* **2019**, *28*, 380–394. [CrossRef]
20. Bilecenoglu, M.; Kaya, M.; Eryigit, A. New Data on the Occurrence of Two Alien Fishes, Pisodonophis Semicinctus and Pomadasys Stridens, from the Eastern Mediterranean Sea. *Mediterr. Mar. Sci.* **2009**, *10*, 151. [CrossRef]
21. Bakshi, M.; Abhilash, P.C. Nanotechnology for Soil Remediation: Revitalizing the Tarnished Resource. In *Nano-Materials as Photocatalysts for Degradation of Environmental Pollutants*; Elsevier: Amsterdam, The Netherlands, 2020; pp. 345–370.
22. Yaqoob, A.A.; Parveen, T.; Umar, K.; Mohamad Ibrahim, M.N. Role of Nanomaterials in the Treatment of Wastewater: A Review. *Water* **2020**, *12*, 495. [CrossRef]
23. Sarkar, A.; Sengupta, S.; Sen, S. Nanoparticles for Soil Remediation. In *Nanoscience and Biotechnology for Environmental Applications*; Springer: Berlin/Heidelberg, Germany, 2019; pp. 249–262.
24. Prasad, R.; Bhattacharyya, A.; Nguyen, Q.D. Nanotechnology in Sustainable Agriculture: Recent Developments, Challenges, and Perspectives. *Front. Microbiol.* **2017**, *8*, 1014. [CrossRef]
25. Alessandrino, L.; Colombani, N.; Aschonitis, V.G.; Mastrocicco, M. Nitrate and Dissolved Organic Carbon Release in Sandy Soils at Different Liquid/Solid Ratios Amended with Graphene and Classical Soil Improvers. *Appl. Sci.* **2022**, *12*, 6220. [CrossRef]
26. Cheng, S.; Chen, T.; Xu, W.; Huang, J.; Jiang, S.; Yan, B. Application Research of Biochar for the Remediation of Soil Heavy Metals Contamination: A Review. *Molecules* **2020**, *25*, 3167. [CrossRef] [PubMed]
27. Yang, X.; Zhang, S.; Ju, M.; Liu, L. Preparation and Modification of Biochar Materials and Their Application in Soil Remediation. *Appl. Sci.* **2019**, *9*, 1365. [CrossRef]
28. Islam, T.; Li, Y.; Cheng, H. Biochars and Engineered Biochars for Water and Soil Remediation: A Review. *Sustainability* **2021**, *13*, 9932. [CrossRef]
29. Liang, M.; Lu, L.; He, H.; Li, J.; Zhu, Z.; Zhu, Y. Applications of Biochar and Modified Biochar in Heavy Metal Contaminated Soil: A Descriptive Review. *Sustainability* **2021**, *13*, 14041. [CrossRef]
30. Song, P.; Xu, D.; Yue, J.; Ma, Y.; Dong, S.; Feng, J. Recent Advances in Soil Remediation Technology for Heavy Metal Contaminated Sites: A Critical Review. *Sci. Total Environ.* **2022**, *838*, 156417. [CrossRef]
31. Xu, D.-M.; Fu, R.-B.; Wang, J.-X.; Shi, Y.-X.; Guo, X.-P. Chemical Stabilization Remediation for Heavy Metals in Contaminated Soils on the Latest Decade: Available Stabilizing Materials and Associated Evaluation Methods-A Critical Review. *J. Clean. Prod.* **2021**, *321*, 128730. [CrossRef]
32. Das, S.; Kim, G.W.; Hwang, H.Y.; Verma, P.P.; Kim, P.J. Cropping with Slag to Address Soil, Environment, and Food Security. *Front. Microbiol.* **2019**, *10*, 1320. [CrossRef]
33. Liyun, Y.; Ping, X.; Maomao, Y.; Hao, B. The Characteristics of Steel Slag and the Effect of Its Application as a Soil Additive on the Removal of Nitrate from Aqueous Solution. *Environ. Sci. Pollut. Res.* **2017**, *24*, 4882–4893. [CrossRef]

34. Wang, X.; Zhu, Y.; Hu, Z.; Zhang, L.; Yang, S.; Ruan, R.; Bai, S.; Tan, H. Characteristics of Ash and Slag from Four Biomass-Fired Power Plants: Ash/Slag Ratio, Unburned Carbon, Leaching of Major and Trace Elements. *Energy Convers. Manag.* **2020**, *214*, 112897. [CrossRef]
35. Tosti, L.; van Zomeren, A.; Pels, J.R.; Dijkstra, J.J.; Comans, R.N.J. Assessment of Biomass Ash Applications in Soil and Cement Mortars. *Chemosphere* **2019**, *223*, 425–437. [CrossRef] [PubMed]
36. Tanji, A.; Benicha, M.; Mrabet, R. Techniques de Production de l'arachide, Résultats d'enquêtes Au Loukkos. *Bull. Transf. Technol. Agric.* **2011**. [CrossRef]
37. Mourabit, F.; Ouassini, A.; Azmani, A.; Mueller, R. Nitrate Occurrence in the Groundwater of the Loukkos Perimeter. *J. Environ. Monit.* **2002**, *4*, 127–130. [CrossRef] [PubMed]
38. El Bakouri, H.; Ouassini, A.; Morillo, J.; Usero, J. Pesticides in Ground Water beneath Loukkos Perimeter, Northwest Morocco. *J. Hydrol.* **2008**, *348*, 270–278. [CrossRef]
39. Sarti, O.; Otal, E.; Morillo, J.; Ouassini, A. Integrated Assessment of Groundwater Quality beneath the Rural Area of R'mel, Northwest of Morocco. *Groundw. Sustain. Dev.* **2021**, *14*, 100620. [CrossRef]
40. Hmamou, M.; Bounakaya, B. The role of artificial impoundments in improving agricultural production in the semi-arid regions of northern morocco. *Geogr. Environ. Sustain.* **2020**, *13*, 32–42. [CrossRef]
41. Metson, A.J. *Methods of Chemical Analysis for Soil Survey Samples*; New Zealand Department of Scientific and Industrial Research: Wellington, New Zealand, 1961.
42. Alam, M.T.; Dai, B.; Wu, X.; Hoadley, A.; Zhang, L. A Critical Review of Ash Slagging Mechanisms and Viscosity Measurement for Low-Rank Coal and Bio-Slags. *Front. Energy* **2021**, *15*, 46–67. [CrossRef]
43. Wang, S.; Jiang, X.M.; Han, X.X.; Wang, H. Fusion Characteristic Study on Seaweed Biomass Ash. *Energy Fuels* **2008**, *22*, 2229–2235. [CrossRef]
44. Zarabi, M.; Jalali, M. Leaching of Nitrogen and Base Cations from Calcareous Soil Amended with Organic Residues. *Environ. Technol.* **2012**, *33*, 1577–1588. [CrossRef]
45. Mengel, K.; Kirkby, E.A.; Kosegarten, H.; Appel, T. Phosphorus. In *Principles of Plant Nutrition*; Springer: Dordrecht, The Netherlands, 2001; pp. 453–479.
46. Jackson, R.S. 5—Site Selection and Climate. In *Wine Science*, 4th ed.; Jackson, R.S., Ed.; Food Science and Technology; Academic Press: San Diego, CA, USA, 2014; pp. 307–346. ISBN 978-0-12-381468-5.
47. Fujii, K.; Hayakawa, C.; Panitkasate, T.; Maskhao, I.; Funakawa, S.; Kosaki, T.; Nawata, E. Acidification and Buffering Mechanisms of Tropical Sandy Soil in Northeast Thailand. *Soil Tillage Res.* **2017**, *165*, 80–87. [CrossRef]
48. Yan, A.; Wang, Y.; Tan, S.N.; Mohd Yusof, M.L.; Ghosh, S.; Chen, Z. Phytoremediation: A Promising Approach for Revegetation of Heavy Metal-Polluted Land. *Front. Plant Sci.* **2020**, *11*, 359. [CrossRef] [PubMed]
49. Remon, E. Tolérance et Accumulation Des Métaux Lourds Par La Végétation Spontanée Des Friches Métallurgiques: Vers de Nouvelles Méthodes de Bio-Dépollution. Ph.D. Thesis, Université Jean Monnet, Saint-Étienne, France, 2006.
50. Stefanowicz, A.M.; Kapusta, P.; Zubek, S.; Stanek, M.; Woch, M.W. Soil Organic Matter Prevails over Heavy Metal Pollution and Vegetation as a Factor Shaping Soil Microbial Communities at Historical Zn–Pb Mining Sites. *Chemosphere* **2020**, *240*, 124922. [CrossRef] [PubMed]
51. Atafar, Z.; Mesdaghinia, A.; Nouri, J.; Homaee, M.; Yunesian, M.; Ahmadimoghaddam, M.; Mahvi, A.H. Effect of Fertilizer Application on Soil Heavy Metal Concentration. *Environ. Monit. Assess.* **2010**, *160*, 83–89. [CrossRef] [PubMed]
52. Kelepertzis, E. Accumulation of Heavy Metals in Agricultural Soils of Mediterranean: Insights from Argolida Basin, Peloponnese, Greece. *Geoderma* **2014**, *221–222*, 82–90. [CrossRef]
53. Sun, C.; Liu, J.; Wang, Y.; Sun, L.; Yu, H. Multivariate and Geostatistical Analyses of the Spatial Distribution and Sources of Heavy Metals in Agricultural Soil in Dehui, Northeast China. *Chemosphere* **2013**, *92*, 517–523. [CrossRef]
54. Mertens, J.; Smolders, E. Zinc. In *Heavy Metals in Soils: Trace Metals and Metalloids in Soils and Their Bioavailability*, 3rd ed.; Alloway, B.J., Ed.; Springer: Berlin, Germanry, 2013; pp. 465–493.
55. Batra, V.S.; Urbonaite, S.; Svensson, G. Characterization of Unburned Carbon in Bagasse Fly Ash. *Fuel* **2008**, *87*, 2972–2976. [CrossRef]
56. Motesharezadeh, B.; Bell, R.W.; Ma, Q. Lime Application Reduces Potassium and Nitrate Leaching on Sandy Soils (Die Kalkanwendung Reduziert Das Auswaschen von Kalium Und Nitrat Auf Sandigen Böden). *J. Kult.* **2021**, *73*, 83–93.
57. Wang, L.; Skjevrak, G.; Hustad, J.E.; Skreiberg, Ø. Investigation of Biomass Ash Sintering Characteristics and the Effect of Additives. *Energy Fuels* **2014**, *28*, 208–218. [CrossRef]
58. Nurmesniemi, H.; Mäkelä, M.; Pöykiö, R.; Manskinen, K.; Dahl, O. Comparison of the Forest Fertilizer Properties of Ash Fractions from Two Power Plants of Pulp and Paper Mills Incinerating Biomass-Based Fuels. *Fuel Process. Technol.* **2012**, *104*, 1–6. [CrossRef]
59. Mandal, J.K.; Mukherjee, S.; Saha, N.; Halder, N.; Biswas, T.; Chakraborty, S.; Hassan, S.; Hassan, M.M.; Abo-Shosha, A.A.; Hossain, A. Assessing the Capability of Chemical Ameliorants to Reduce the Bioavailability of Heavy Metals in Bulk Fly Ash Contaminated Soil. *Molecules* **2021**, *26*, 7019. [CrossRef]
60. Querol, X.; Alastuey, A.; Moreno, N.; Alvarez-Ayuso, E.; García-Sánchez, A.; Cama, J.; Ayora, C.; Simón, M. Immobilization of Heavy Metals in Polluted Soils by the Addition of Zeolitic Material Synthesized from Coal Fly Ash. *Chemosphere* **2006**, *62*, 171–180. [CrossRef] [PubMed]

61. Dermatas, D.; Meng, X. Utilization of Fly Ash for Stabilization/Solidification of Heavy Metal Contaminated Soils. *Eng. Geol.* **2003**, *70*, 377–394. [CrossRef]
62. Ciccu, R.; Ghiani, M.; Peretti, R.; Serci, A.; Zucca, A. Heavy Metal Immobilization Using Fly Ash in Soils Contaminated by Mine Activity. In Proceedings of the 2001 International Ash Utilization Symposium, Lexington, KY, USA, 22–24 October 2001.
63. Hu, X.; Huang, X.; Zhao, H.; Liu, F.; Wang, L.; Zhao, X.; Gao, P.; Li, X.; Ji, P. Possibility of Using Modified Fly Ash and Organic Fertilizers for Remediation of Heavy-Metal-Contaminated Soils. *J. Clean. Prod.* **2021**, *284*, 124713. [CrossRef]
64. Ciccu, R.; Ghiani, M.; Serci, A.; Fadda, S.; Peretti, R.; Zucca, A. Heavy Metal Immobilization in the Mining-Contaminated Soils Using Various Industrial Wastes. *Miner. Eng.* **2003**, *16*, 187–192. [CrossRef]
65. Ou, Y.; Ma, S.; Zhou, X.; Wang, X.; Shi, J.; Zhang, Y. The Effect of a Fly Ash-Based Soil Conditioner on Corn and Wheat Yield and Risk Analysis of Heavy Metal Contamination. *Sustainability* **2020**, *12*, 7281. [CrossRef]
66. Chong, I.Q.; Azman, E.A.; Ng, J.F.; Ismail, R.; Awang, A.; Hasbullah, N.A.; Murdad, R.; Ahmed, O.H.; Musah, A.A.; Alam, M.A. Improving Selected Chemical Properties of a Paddy Soil in Sabah Amended with Calcium Silicate: A Laboratory Incubation Study. *Sustainability* **2022**, *14*, 13214. [CrossRef]
67. Bruand, A.; Hartmann, C.; Lesturgez, G. Physical Properties of Tropical Sandy Soils: A Large Range of Behaviours. In *Management of Tropical Sandy Soils for Sustainable Agriculture*; A Holistic Approach for Sustainable Development of Problem Soils in the Tropics; FAO: Rome, Italy, 2005.
68. Matichenkov, V.V.; Bocharnikova, E.A.; Calvert, D.V.; Snyder, G.H. Comparison Study of Soil Silicon Status in Sandy Soils of South Florida. In Proceedings of the Fifty-Ninth Annual Meeting of the Soil and Crop Science Society of Florida, Sarasota, FL, USA, 22–24 September 1999; Volume 59, pp. 132–137.

**Disclaimer/Publisher's Note:** The statements, opinions and data contained in all publications are solely those of the individual author(s) and contributor(s) and not of MDPI and/or the editor(s). MDPI and/or the editor(s) disclaim responsibility for any injury to people or property resulting from any ideas, methods, instructions or products referred to in the content.

Article

# Assessing Soil Organic Carbon Stocks and Particle-Size Fractions across Cropping Systems in the Kiti Sub-Watershed in Central Benin

Arcadius Martinien Agassin Ahogle [1,2,*], Felix Kouelo Alladassi [1], Tobi Moriaque Akplo [1], Hessou Anastase Azontonde [3] and Pascal Houngnandan [1,4]

[1] Laboratoire de Microbiologie des Sols et d'Ecologie Microbienne, Faculté des Sciences Agronomiques, Université d'Abomey-Calavi, Abomey-Calavi BP 711, Benin
[2] Department of Spatial and Environmental Planning, Kenyatta University, Nairobi P.O. Box 43844-00100, Kenya
[3] Institut National des Recherches Agricoles du Bénin, CRA-Agonkanmey, Cotonou BP 884, Benin
[4] Ecole de Gestion et de Production Végétale et Semencière, Université Nationale d'Agriculture, Ketou BP 43, Benin
* Correspondence: ahoglearcadius@gmail.com

**Citation:** Ahogle, A.M.A.; Alladassi, F.K.; Akplo, T.M.; Azontonde, H.A.; Houngnandan, P. Assessing Soil Organic Carbon Stocks and Particle-Size Fractions across Cropping Systems in the Kiti Sub-Watershed in Central Benin. *C* 2022, *8*, 67. https://doi.org/10.3390/c8040067

Academic Editors: Indra Pulidindi, Pankaj Sharma and Aharon Gedanken

Received: 25 October 2022
Accepted: 20 November 2022
Published: 23 November 2022

**Publisher's Note:** MDPI stays neutral with regard to jurisdictional claims in published maps and institutional affiliations.

**Copyright:** © 2022 by the authors. Licensee MDPI, Basel, Switzerland. This article is an open access article distributed under the terms and conditions of the Creative Commons Attribution (CC BY) license (https://creativecommons.org/licenses/by/4.0/).

**Abstract:** Soil organic carbon storage in agricultural soil constitutes a crucial potential for sustainable agricultural productivity and climate change mitigation. This paper aimed at assessing soil organic carbon stock and its distribution in three particle size fractions across five cropping systems located in Kiti sub-watershed in Benin. Soil samples were collected using a grid sampling method on four soil depth layers: 0–10, 10–20, 20–30 and 30–40 cm in five cropping systems maize–cotton relay cropping (MCRC), yam–maize intercropping (YMI), teak plantation (TP), 5-year fallow (5YF) and above 10-year fallow (Ab10YF) from July to August 2017. Soil organic carbon stock (C stock) was estimated for the different soil layers and particle-size fractionation of soil organic matter was performed considering three fractions. The fractions coarse particulate organic matter (cPOM: 250–2000 µm), fine particulate organic matter (fPOM: 53–250 µm) and non-particulate organic matter (NOM: <53 µm) were separated from two soil depth layers: 0–10 and 10–20 cm. The results showed that fallow lands Ab10YF and 5YF exhibited the highest C stock, 22.20 and 17.74 Mg C·ha$^{-1}$, while cultivated land under tillage MCRC depicted the lowest, C stock 11.48 Mg C·ha$^{-1}$. The three organic carbon fractions showed a significant variation across the cropping systems with the NOM fraction holding the largest contribution to total soil organic carbon for all the cropping systems, ranging between 3.40 and 7.99 g/kg. The cPOM and fPOM were the most influenced by cropping systems with the highest concentration observed in Ab10YF and 5YF. The findings provide insights for upscaling farm management practices towards sustainable agricultural systems with substantial potential for carbon sequestration and climate change mitigation.

**Keywords:** carbon sequestration; sustainable farming systems; particulate organic carbon; particle-size fractionation

## 1. Introduction

Feeding an ever-growing population, expected to pass nine billion by 2050, in the tight context of climate change and resource degradation sets the greatest challenge for agricultural systems in the world [1]. Food and nutrition security is challenging in sub-Saharan Africa (SSA) where agricultural production is mainly rain-fed, relying mostly on traditional modes of farming with low input, i.e., most farmers do not use improved seed varieties and irrigation systems, apply low-rate organic amendment and mineral fertilizers, export crop residues and do not implement mechanized farming operations [2,3]. This type of farming system often leads to soil fertility depletion and disruption of its biochemical processes [4]. Along with the continuous exploitation of soil without replenishment, more pressure has been imposed on the soil to cope with the ever-increasing food demand [5]. This has resulted in increasing soil degradation and subsequently a decrease in agricultural production [6].

Soils are crucial resources for agricultural production in that they help in water filtering, biodiversity preservation, atmospheric carbon storage and host biogeochemical processes [7,8]. As an essential reservoir of atmospheric carbon, soils have a vital role in the mitigation of greenhouse gas emissions [9]. Soil management is vital in environmental sustainability and achievement of the sustainable development goals (SDGs).

In SSA, soil fertility decline is a significant constraint hindering agricultural production [10]. In Benin, a country located in West Africa, about 70% of the total arable lands has been classified as low to very-low fertility [11]. This low soil fertility level is partially attributed to the intrinsic properties of these soils, i.e., low soil organic carbon (SOC) content and low-cation-exchange capacity [12]. In addition, poor farming practices, such as burning, crop residues exportation and nutrient miming contribute further to soil fertility depletion [4,12]. Indeed, soil organic carbon constitutes a key component of soil fertility and agroecosystem sustainability [13]. The literature has shown that the high content of soil organic matter in the surface layer is significantly correlated with a lower susceptibility to water erosion in West Africa [14–17]. According to Paul et al. [18], soil aggregation and soil structure stability increase with soil organic carbon content. Similarly, soil fauna diversity and activity are directly related to soil organic carbon content [19]. Furthermore, soil organic carbon influences fertilizer efficiency in agricultural production. However, different pools of SOC are involved in these processes [20]. The labile pool (<53–2000 μm) which has a few days to months turnover stimulates microbial activity and nutrient cycling [21], while the non-labile (<53 μm) or recalcitrant pool which has a long turnover is responsible of carbon sequestration and climate regulation [22]. Soil fertility in terms of nutrient availability is sensitive to the labile pools of SOC, whereas soil potentiality for climate regulation through carbon sequestration and ecosystem sustainability depends much more on the non-labile pool of SOC [23]. Furthermore, discussions by the United Nations Framework Convention on Climate Change (UNFCCC) and other international fora placed agricultural soil in a vital position for mitigating climate change [24,25]. Hence, soil management practices that promote carbon storage are essential for sustainable production systems and climate change adaptation and mitigation. Therefore, understanding the dynamics of soil organic carbon stock and its subsequent pools in different cropping systems is essential for designing and implementing sustainable farming systems [26].

Cropping system specifications and farm management practices have great impact on carbon storage and its spatiotemporal kinetics [27,28]. Soil potential for organic matter storage depends on soil type, soil management practices and climate conditions [2,29,30]. Precipitation positively affects SOC content, while temperature adversely affects SOC vertical distribution [31]. Previous studies [29,32–38] have reported that farm management practices such as crop residues restitution, mulching, organic amendment, cover crop, legume intercropping and biochar application to soil have positive effects on soil organic carbon storage. However, the quality of the organic resources used is a key parameter of carbon storage in the soil. Choudhury et al. [39] and Yoo and Wander [40] demonstrated that tillage leads to soil aggregate break-up and soil organic carbon mineralization, while no tillage induces higher soil particle aggregation, carbon sequestration and particulate organic matter (POM) buildup [41].

In Benin, studies evaluating the effect of cropping systems on SOC are limited to a few studies reporting on agroforestry systems [42,43], cereal–legume-based cropping systems [14,44], palm oil-based cropping systems [33,45,46], vegetable cropping systems under poultry and sheep dung manures and fallow land [32,42]. Despites that, the dynamics of soil organic carbon stock (C stock) and its pools are at the center of various discussions on climate and sustainable development, with more research interest over the last decades in Benin, very few studies have reported on smallholder farming systems which are very complex in terms of resource endowment and integration with various spatiotemporal arrangements [47,48]. Knowledge related to the effects of different farming systems on the C stock in the region is still unclear and limited to a few research studies [32,33,42,43,45,46,49]. Therefore, the objectives of this study were to (i) investigate the C stock across selected cropping systems and (ii) assess the particle-size distribution of SOC in these cropping systems at a watershed scale. We hypothesized that fallow land and teak plantation store more carbon than cultivated lands and the POM fraction is more sensitive to cropping system characteristics than the NOM.

## 2. Materials and Methods

### 2.1. Study Area

The study was carried out in Kiti sub-watershed in central Benin. The sub-watershed is part of the Zou watershed which is one of the biggest watersheds of Benin. The sub-watershed of Kiti lies between 2°4′00″–2°12′00″ longitude East and 7°20′00″–7°29′00″ latitude North and covers an area of 85,690.8 ha (Figure 1). The mainstream of the sub-watershed is Kiti which is a tributary of the Zou River. The climate in this area is tropical Sudano-Guinean, with a bimodal rainfall pattern. Daily temperatures range from 26 to 31 °C, and annual rainfall averages range between 1000 to 1200 mm [43]. Soils are primarily ferruginous tropical soil with concretions [50] classified as Luvisols [51]. These soils are characterized by a yellowish to light brown sandy horizon on brownish red clay, very concretionary with angular quartz gravels and occasionally ferruginous [52]. They have substantial alterations with an accumulation of ferric hydrates associated with very little oxidized aluminum [52]. The texture is sandy-clay with poor drainage at deeper layers due to high clay eluviation from the surface layer [11]. The sub-watershed of Kiti is part of the Central Benin cotton agroecological zone (ZAE 5). The vegetation is a lightly wooded savannah with sparse shrubs of natural trees and small-sized plantations, with agriculture being the predominant livelihood means for the communities around the watershed. This sub-watershed was purposely chosen because it is an area of intensive agricultural production of cash crop (e.g., cotton) and staple food crops (e.g., maize) in the Zou watershed with a substantial impact on smallholder farmers' livelihoods.

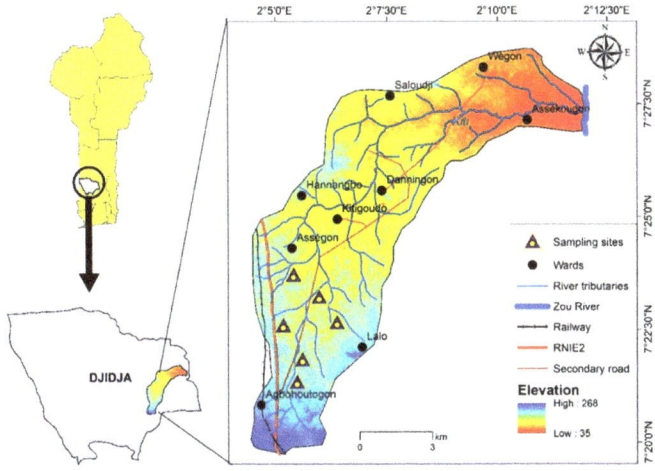

**Figure 1.** Map showing the sampling sites.

### 2.2. Cropping Systems

As the primary potential characteristic of this agroecological zone, cotton (*Gossypium* spp.) is the main cash crop cultivated in this region. The cropping systems are dominated by maize (*Zea mays* L.) and cotton-based cropping systems. Maize–cotton relay cropping (MCRC) is the primary cropping system implemented in the watershed. The maize–cotton system is a relay cropping system characterized by manual ploughing at a maximum depth of 20 cm. Maize is sown at the beginning of the long rainy season (between mid-March and mid-April), while cotton is sown in the maize cob maturity stages (between 15 July and 30 August). The use of mineral fertilizer is globally low for maize and other crops, while 150 kg/ha of NPK (15-15-15) and 50 kg/ha of urea (46% N) is usually applied for cotton. Yam (*Dioscorea* spp.) is grown in the watershed in small plots. As yam cropping requires high soil fertility, it is generally cultivated at the top of the crop rotation on new fallow land. Thus, yam producers constantly look for new fallow or forest lands to convert into farmland [53]. The yam–maize intercropping system (YMI) is characterized by mound ploughing at about 40 cm high. In the sub-watershed of Kiti, yam is intercropped with maize with low or no fertilizer input. In the cropping systems MCRC and YMI, maize and yam residues are spread in the furrows while cotton stalks are gathered and burned for pest management. Although, most of the farmers in the watershed leave the crop residues on the farm as mulch for replenishing soil fertility, these crop residues are commonly grazed by livestock belonging to transhumant pastoralists passing through the region in search of graze for their livestock, especially during the dry season [54–56]. This leads to almost a complete exportation of crop residue from the farm, leaving bare the soil surface, which becomes more susceptible to erosion from heavy winds during the dry season and rain at the beginning of the wet period [57]. The lands which have higher proportion of soil concretion are difficult to plough and are generally used for tree plantation, notably teak plantation (*Tectona grandis*). The teak plantation investigated in this study is a plantation established since 1998. The wooded trees are sold for use as posts or poles with a diameter of 5 to 15 cm and an average harvesting period ranging from 5 to 10 years [58]. Although the fallow period has generally reduced due to land shortage, some farmers in the area still observe a fallow period ranging from 5 to 10 years and above. During farmland exploration, two typical fallow lands have been identified: 5-year fallow (5YF) and above 10-year fallow (Ab10YF). The fallow lands, 5YF and Ab10YF, were covered by natural vegetation and shrubs, including *Vitellaria paradoxa*, *Azadirachta indica*, *Nauclea latifolia*, *Danielia oliveri*, *Imperata cylindrica* and *Cleome viscosa*. However, these fallow lands are influence by seasonal vegetation fires during the dry season [59,60].

*2.3. Sample Collection and Analysis*

The study used an experimental research design considering the cropping systems as the principal factor to investigate. The cropping systems included maize–cotton relay cropping (MCRC), yam–maize intercropping (YMI), teak plantation (TP), five-year fallow (5YF) and above ten-year fallow (10YF) (Table 1). Soil samples were collected from July to August 2017, using a grid establishment approach [61]. The grids were constructed using a step of 20 m × 30 m on a total area of 6 sq.km covering the five cropping systems investigated. The grids for soil sample collection were selected randomly in each cropping system. A total of 50 grids were sampled: 18 grids in MCRC, 8 in YMI, 8 for TP, 8 in 5YF and 8 in Ab10YF. The high number of sampling grids in MCRC compared to the others was because of the high coverage of this cropping system across the sub-watershed. Soil samples were collected at four depths in each grid: 0–10, 10–20, 20–30 and 30–40 cm. The collected soil samples were air-dried for three weeks, mechanically crushed using a stainless-steel roller and sieved through a 2 mm sieve for laboratory analyses. Soil organic carbon content in soil was determined using boiled potation bichromate in acidic conditions, as described in Okalebo et al. [62]. The absorbance of the samples was read with a spectrophotometer at a wavelength of 600 nm. Furthermore, the soil pH was determined in a distilled water ratio of 1:2.5 and the Robinson pipette method was used for soil texture determination [63]. In each sampled grid, the cylinder method (calibrated density cylinders of a known volume of 100 cm$^3$) was used to collect the samples for soil bulk-density determination for each layer. The contents of the cylinder were weighed after drying at 105 °C in an oven for 24 h. The bulk density BD is given by the ratio of dry weight to volume.

Table 1. Cropping systems history.

| Years | Cropping Systems | | | | |
|---|---|---|---|---|---|
| | MCRC | YMI | TP | 5YF | Ab10YF |
| 2005–2006 | Fallow | Maize–cotton | Teak plantation | Maize–cotton | Fallow |
| 2006–2007 | Fallow | Maize–cotton | Teak plantation | Maize–cotton | Fallow |
| 2007–2008 | Maize–soybean | Maize–soybean | Teak plantation | Maize–cotton | Fallow |
| 2008–2009 | Maize–soybean | Maize–soybean | Teak plantation | Maize–cotton | Fallow |
| 2009–2010 | Maize–soybean | Fallow | Teak plantation | Maize–cotton | Fallow |
| 2010–2011 | Maize–cotton | Fallow | Teak plantation | Maize–cotton | Fallow |
| 2011–2012 | Maize–cotton | Fallow | Teak plantation | Maize–cotton | Fallow |
| 2012–2013 | Maize–cotton | Fallow | Teak plantation | Fallow | Fallow |
| 2013–2014 | Maize–cotton | Fallow | Teak plantation | Fallow | Fallow |
| 2014–2015 | Maize–cotton | Fallow | Teak plantation | Fallow | Fallow |
| 2015–2016 | Maize–cotton | Yam–maize | Teak plantation | Fallow | Fallow |
| 2016–2017 | Maize–cotton | Yam–maize | Teak plantation | Fallow | Fallow |

### 2.4. Soil Carbon Stock Calculation

Soil organic carbon stock computation was based on soil bulk-density, the thickness of the soil layer and the proportion of fine soil. The C stock was computed using Equation (1) [42]. Carbon stock was estimated for the four soil layers 0–10, 10–20, 20–30 and 30–40 cm. C stock was considered for each layer, for 0–30 cm and for 0–40 cm.

$$\text{C stock} = \sum_{Depth=1}^{Depth=n} \text{C stock}_{Depth} = \sum_{Depth=1}^{Depth=n} (\text{SOC} \times \text{BD} \times P \times (1 - \text{frag}) \times 10) \quad (1)$$

where C stock (Mg C·ha$^{-1}$) is the sum of soil organic carbon stock for the different layers considered, C stock$_{Depth}$ (Mg C·ha$^{-1}$) is the stock of organic carbon at a specific soil depth, SOC (g·C kg$^{-1}$) is the concentration of soil total organic carbon, BD (g·cm$^{-3}$) is soil bulk-density, P (m) is the thickness of the soil layer, frag is the percentage volume of coarse fragments/100 and n is the number of layer considered.

### 2.5. Soil Organic Matter Particle-Size Fractionation

There are various methods for soil organic carbon partitioning into its granulometric functional pools. The particle-size fractionation of soil organic carbon allows to assess the distribution of SOC according to particle sizes. It gives useful information on the proportion of the different types of soil organic matter, their chemical composition, and potential dynamics in soil which are essential in evaluating the sustainability of various land management options for soil organic carbon rehabilitation. Soil organic carbon fractionation used in this study was adapted from the method develop by Feller [64] for coarse texture and poor humus content soil, used with good overall accuracy by Sainepo et al. [65] in Kenya; by Koussihouèdé et al. [32] in Benin and Gura et al. [66] in South Africa.

A reciprocal shaker was used to mix 50 g of soil with 300 mL of distilled water with 10 mL of Calgon solution (10% sodium hexametaphosphate, 50 g·L$^{-1}$) for 15 h. The solution was passed through a series of nested sieves of sizes 2000 µm, 250 µm and 53 µm in a wet sieving apparatus with deionized water. The particles that passed through the 53 µm was referred as the non-particulate organic matter fraction (NOM). The fraction 53–250 µm was referred as the fine particulate organic matter fraction (fPOM), while the 250–2000 µm fraction was considered as the coarse particulate organic fraction (cPOM) [67]. In the cPOM fraction, plant materials such as plant residues and roots that had partially broken down were carefully separated. The isolated particles for cPOM and fPOM were washed with deionized water until clean and backwashed into an evaporation dish. The fraction that passed through the 53 µm sieve was collected in a volumetric flask, quantified, thoroughly homogenized and a sample of 100 mL was collected in an evaporation dish. The evaporation dishes were dried at 65 °C till a constant weight. The oven-dried soil particles were weighed and placed in dry porcelain crucibles and heated in a muffle furnace at 450 °C for 4 h to separate the mineral particles from the organic particles. After cooling, the organic matter contained in each fraction was determined as shown in Equation (2).

$$\text{Fraction} = \frac{\text{Weight at 65 °C} - \text{Weight at 450 °C}}{\text{Weight at 65 °C}} \quad (2)$$

The organic carbon content in the fractions were calculated using the coefficient 1.724 considering that organic matter comprises 58% carbon [68]. Particle-size fractionation of SOC was caried out for two soil layers, 0–10 and 10–20 cm. For each sample the percentage recovery was calculated by the ratio of the sum of the weight of the three fractions by the initial weight of 50 g, multiplied by 100 (Equation (3)) and the total fraction was reported considering 1000 g of soil. The enrichment factor (EF) in each of the fractions was calculated according to Equation (4) [69].

$$\% \text{ Recovery} = \frac{\text{Weigth cPOM} + \text{WeigthfPOM} + \text{WeigthNOM}}{50} \times 100 \quad (3)$$

$$EF = \frac{\text{SOC fraction (g/kg)}}{\text{totalSOC (g/kg)}} \times 100 \quad (4)$$

### 2.6. Statistical Analysis

The dataset was screened for normality and variance homogeneity using the Shapiro–Wilk test [70] and Bartlett's test [71]. Thereafter, a one-way ANOVA was used to determine significant differences among the different cropping systems. Tukey's post hoc test was used to separate the means, in case of significant difference at a 5% significance threshold level. All analyses were performed in R software version 4.1.

## 3. Results

### 3.1. Soil Properties

In this study, there was no significant difference in soil texture (p = 0.09) across the cropping systems and soil was sandy-clay loam or sandy-clay depending on their silt content (Table 2). Soil pH varied significantly ($p = 0.03$) between the cropping systems and soil under MCRC exhibited the lowest pH values. The soil under MCRC was classified as acidic while the soil under other cropping systems were classified as slightly acidic. Soil bulk-density (BD) values were significantly higher for the surface depth layers of 0–10 and 10–20 cm in the soil under fallow (5YF and Ab10YF) and teak plantation than the soil under cropping (MCRC and YMI). However, for the sub-layer depths (20–30 and 30–40), regardless of the cropping system, BD values increased with YMI showing the lowest BD while MCRC, TP, 5YF and Ab10YF exhibited the highest values.

Table 2. Soil properties under the different cropping systems.

| Soil Properties | Soil Depth | MCRC | Yam-Maize | TP | 5YF | Ab10YF | p-Value |
|---|---|---|---|---|---|---|---|
| Clay (g·kg$^{-1}$) | 0–20 | 333.13 ± 38 a | 389.67 ± 52 a | 344.13 ± 45 a | 366.36 ± 48 a | 379.94 ± 85 a | 0.090 ns |
| Silt (g·kg$^{-1}$) | 0–20 | 31.90 ± 36 a | 27.20 ± 44 a | 28.90 ± 30 a | 39.59 ± 66 a | 39.67 ± 52 a | 0.310 ns |
| Sand (g·kg$^{-1}$) | 0–20 | 636.80 ± 69 a | 599.77 ± 63 a | 624.80 ± 57 a | 595.32 ± 75 a | 583.47 ± 86 a | 0.100 ns |
| Soil texture | 0–20 | Sandy clay loam | Sandy clay | Sandy clay loam | Sandy clay | Sandy clay | - |
| pH | 0–20 | 5.80 ± 0.2 b | 6.05 ± 0.3 ab | 6.1 ± 0.1 ab | 6.2 ± 0.2 a | 6.3 ± 0.1 a | 0.010 * |
| BD | 0–10 | 1.27 ± 0.17 ab | 1.41 ± 0.27 a | 1.14 ± 0.08 b | 1.43 ± 0.21 a | 1.11 ± 0.18 b | 0.024 * |
| | 10–20 | 1.46 ± 0.16 b | 1.55 ± 0.22 b | 1.61 ± 0.09 ab | 1.54 ± 0.08 ab | 1.68 ± 0.01 a | 0.009 * |
| | 20–30 | 1.67 ± 0.12 a | 1.47 ± 1.14 b | 1.71 ± 015 a | 167 ± 0.21 a | 1.69 ± 0.013 a | 0.032 * |
| | 30–40 | 1.73 ± 0.2 a | 1.61 ± 0.2 b | 1.67 ± 0.13 a | 1.68 ± 0.18 a | 1.77 ± 0.08 a | 0.040 * |

MCRC: maize–cotton relay cropping; 5YF: 5-year fallow; Ab10YF: above 10-year fallow; YMI: yam–maize intercropping system; TP: teak plantation; BD: bulk density. Means that do not share a letter are significantly different at α = 0.05. ns non-significant at 5%; * p value significant at 5%.

*3.2. Total Soil Organic Carbon Content (SOC) and Soil Organic Carbon Stock (C Stock) across the Cropping Systems*

The variations in SOC and C stock from the four layers, 0–10, 10–20, 20–30 and 30–40 cm are presented in Table 3. For the soil layer 0–10 cm, the SOC content significantly varied ($p < 0.001$) between the cropping systems and ranged from 3.14–24.1 g C·kg$^{-1}$. The highest SOC was recorded with Ab10YF while the lowest was recorded with MCRC. In the layer 10–20 cm, the SOC content showed significant ($p = 0.011$) differences between the cropping systems with 5YF and YMI recording the highest and the lowest SOC, respectively (5.22 and 2.21 g C·kg$^{-1}$). At the soil layer of 20–30 cm, the SOC content between the cropping systems was not significantly different ($p = 0.1$) with values ranging between 2.10–2.84 mg C·kg$^{-1}$. In the soil layer of 30–40 cm, the SOC showed significant variation ($p = 0.021$) with values ranging from 1.36–4.43 g C·kg$^{-1}$. The cropping systems YMI and 5YF exhibited the lowest and the highest concentrations of SOC, 1.36 and 4.43 g C·kg$^{-1}$, respectively.

**Table 3.** Soil organic carbon content and carbon stock per soil depth layer.

| Soil Properties | Depth (cm) | Cropping Systems | | | | | p-Value |
|---|---|---|---|---|---|---|---|
| | | MCRC | YMI | TP | 5YF | Ab10YF | |
| SOC (g·kg$^{-1}$) | 0–10 | 3.14 ± 0.98 b | 7.37 ± 4.24 b | 5.64 ± 2.62 b | 4.94 ± 2.3 b | 24.1 ± 11.6 a | <0.001 *** |
| | 10–20 | 3.03 ± 1.33 b | 2.21 ± 0.47 b | 3.16 ± 1.04 ab | 5.22 ± 3.78 a | 2.43 ± 0.23 b | 0.011 * |
| | 20–30 | 2.55 ± 0.79 a | 2.06 ± 0.55 a | 2.84 ± 1.0 a | 2.83 ± 1.04 a | 2.10 ± 0.43 a | 0.1 ns |
| | 30–40 | 2.24 ± 0.61 ab | 1.36 ± 0.29 b | 2.64 ± 0.85 ab | 4.43 ± 4.37 a | 2.27 ± 2.47 ab | 0.021 * |
| C stock (Mg C·ha$^{-1}$) | 0–10 | 3.25 ± 0.52 c | 7.34 ± 3.64 bc | 4.72 ± 1.9 bc | 5.73 ± 2.35 bc | 18.1 ± 6.35 a | <0.001 *** |
| | 10–20 | 3.0 ± 0.7 ab | 2.98 ± 0.63 ab | 3.31 ± 2.05 ab | 6.29 ± 5.0 a | 1.39 ± 0.5 b | 0.001 ** |
| | 20–30 | 2.90 ± 1.07 a | 2.90 ± 1.07 a | 3.08 ± 1.13 a | 2.47 ± 0.91 ab | 1.35 ± 0.1 b | 0.006 ** |
| | 30–40 | 2.24 ± 0.49 ab | 1.81 ± 0.57 ab | 1.97 ± 0.96 ab | 3.26 ± 2.26 a | 1.35 ± 0.57 b | 0.012 * |

MCRC: maize–cotton relay cropping; 5YF: 5-year fallow; Ab10YF: above 10-year fallow; YMI: yam–maize intercropping system; TP: teak plantation; SOC: soil organic carbon; C stock: carbon organic stock. Means that do not share a letter are significantly different at α = 0.05; * p value significant at 5%; ** p value significant at 1%; *** p value significant at 0.1%. ns: non-significant at 5%

The C stock showed significant variation between the cropping systems and soil depth layers ($p < 0.001$, $p = 0.001$, $p = 0.006$, $p = 0.012$). Regardless of the cropping systems, C stocks were higher for the surface layer 0–10 cm and showed a decreasing trend towards sub-surface layers. For the soil layer 0–10 cm, Ab10YF recorded the highest C stock (18.1 Mg C·ha$^{-1}$), while MCRC recorded the lowest (3.25 Mg C·ha$^{-1}$). In the layer of 10–20 cm, 5YF showed the highest C stock (6.29 Mg C·ha$^{-1}$), while Ab10YF recorded the lowest C stock 1.39 Mg C·ha$^{-1}$ and the cropping systems ranged in the order 5YF > TP > MCRC > YMI > Ab10YF. In the layer of 20–30 cm, C stock in the soil ranged from 1.35 to 3.26 Mg C·ha$^{-1}$ and Ab10YF and 5YF exhibited the lowest and the highest C stocks, respectively. In the layer of 30–40 cm, the C stock ranged between 3.26 and 1.35 Mg C·ha$^{-1}$ with Ab10YF and 5YF recording the lowest C stock values. In addition, there was a significant ($p = 0.012$ and $p = 0.02$) difference between the cropping systems for the C stock at the layer of 0–30 cm and at the layer 0–40 cm (Figure 2). Considering the total C stock of the layer 0–40 cm, the cropping systems Ab10YF exhibited the highest C stock (22.20 Mg C·ha$^{-1}$) while MCRC recorded the lowest (10.31 Mg C·ha$^{-1}$). Considering the total C stock for the surface layer of 0–30 cm (Figure 2), the highest C stock was recorded at Ab10YF and the lowest at MCRC 20.84 and 9.23 Mg C·ha$^{-1}$, respectively.

*3.3. Organic Carbon Concentrations in Particle Size Fractions*

The average total fraction masses ranged between 969.33 and 991.84 mg·frac·g$^{-1}$ soil indicating an average recovery rate varying between 96.9 and 99.1% (Table 4). Regardless of the cropping system and the soil layer, the non-particulate organic fractions (NOM) showed the highest carbon concentration 3.40–7.99 mg·kg$^{-1}$. The fine particulate organic fractions (fPOM) and the coarse particulate fractions (cPOM) depicted the lowest carbon content between 0.56 and 2.3 mg/kg. In the layer 0–10 cm, carbon concentration in the cropping systems Ab10YF and 5YF were significantly higher in the three fractions NOM ($p = 0.003$), fPOM ($p = 0.01$) and cPOM ($p = 0.02$). The cropping systems YMI and MCRC recorded the lowest carbon concentrations. For the layer 10–20 cm, 5YF exhibited the highest carbon content in NOM, while MCRC and YMI showed the lowest (1.05 and 1.26 mg·kg$^{-1}$, respectively). For this layer, SOC concentration in the NOM fraction varied significantly between the cropping systems ($p = 0.04$). No significant difference was observed for the carbon concentration in the fPOM fraction. However, the carbon content in the cPOM fraction varied significantly ($p = 0.04$) between the cropping systems, with Ab10YF and 5YF and TP recording the highest carbon concentrations 1.24, 1.12 and 0.84 g/kg, respectively.

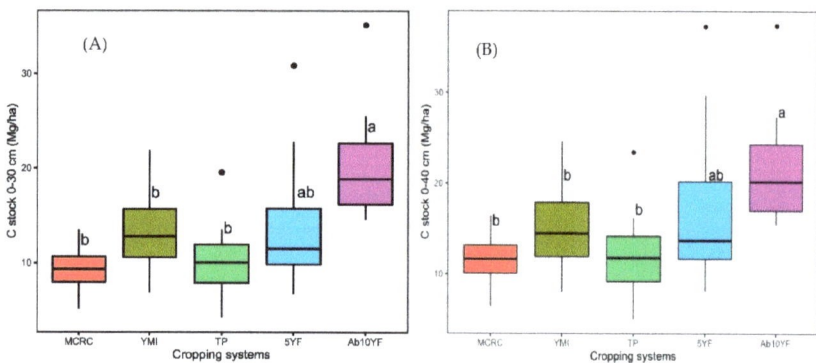

**Figure 2.** Total carbon stock for soil layers 0–30 cm (**A**) and 0–40 cm (**B**) MCRC: maize–cotton relay cropping; 5YF: 5-year fallow; Ab10YF: above 10-year fallow; YMI: yam–maize intercropping system; TP: teak plantation. Means that do not share a letter are significantly different at α = 0.05.

**Table 4.** Carbon concentrations in particle size fractions across the cropping systems for 0–10 and 10–20 cm.

| Depth (cm) | Fractions | MCRC (g·kg$^{-1}$) | YMI (g·kg$^{-1}$) | TP (g·kg$^{-1}$) | 5YF (g·kg$^{-1}$) | Ab10YF (g·kg$^{-1}$) | $p$ Value |
|---|---|---|---|---|---|---|---|
| 0–10 | NOM | 3.40 ± 0.4 d | 4.46 ± 0.38 c | 3.73 ± 0.33 d | 5.93 ± 0.39 b | 7.99 ± 0.21 a | 0.003 ** |
|  | fPOM | 1.19 ± 0.21 b | 1.23 ± 0.32 b | 1.28 ± 0.3 b | 1.89 ± 0.2 ab | 2.24 ± 0.12 a | 0.01 * |
|  | cPOM | 0.70 ± 0.4 b | 0.71 ± 0.42 b | 1.08 ± 0.9 ab | 2.09 ± 0.8 a | 2.3 ± 5.33 a | 0.02 * |
| Total fraction mass (g·kg$^{-1}$ soil) |  | 986.62 | 978.67 | 971.22 | 981.45 | 969.33 |  |
| 10–20 | NOM | 1.05 ± 0.38 c | 1.26 ± 0.25 c | 1.66 ± 0.30 b | 2.53 ± 0.27 a | 1.81 ± 0.99 b | 0.04 * |
|  | fPOM | 1.14 ± 0.47 a | 0.99 ± 0.29 a | 0.78 ± 0.21 a | 0.92 ± 0.29 a | 0.98 ± 0.33 b | 0.044 * |
|  | cPOM | 0.56 ± 0.11 b | 0.41 ± 0.21 b | 0.84 ± 0.32 ab | 1.12 ± 0.17 a | 1.24 ± 0.13 a | 0.04 * |
| Total fraction mass (g/kg soil) |  | 991.84 | 988.33 | 975.52 | 978.56 | 989.42 |  |

MCRC: maize–cotton relay cropping; 5YF: 5-year fallow; Ab10YF: above 10-year fallow; YMI: yam–maize intercropping system; TP: teak plantation; fPOM: fine particulate organic matter; cPOM: coarse particulate organic fraction; NOM: non-particulate organic matter. Means that do not share a letter are significantly different at α = 0.05; * $p$ value significant at 5%; ** $p$ value significant at 1%.

### 3.4. Carbon Enrichment Factor (EF) in Particle-Size Fractionation

The contribution of each particle-size organic matter fraction to the total organic carbon content expressed using the enrichment factor for the two layers revealed that regardless of the cropping systems and the layer, the NOM fraction exhibited the greatest contribution to the total SOC, while cPOM exhibited the lowest contribution (Figure 3). In the two layers, MCRC recorded the highest contribution of the NOM fraction 71.9% and 75.03%. The cropping systems Ab10YF, YMI and TP recorded the highest carbon contribution from the cPOM between 7.1% and 22% for the two layers and between 28.89% and 44.41% for fPOM. In addition, EF values showed that the cPOM and fPOM were the most influenced by the cropping systems.

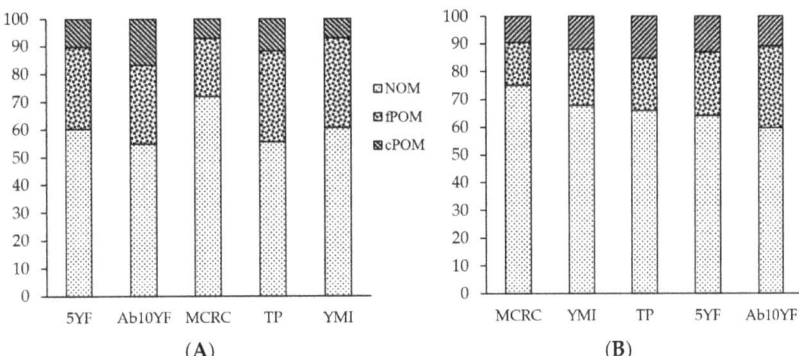

**Figure 3.** Carbon enrichment factor EF of the organic matter fractions (**A**) 0–10 cm layer, (**B**) 10–20 cm layer. MCRC: maize–cotton relay cropping; 5YF: 5-year fallow; Ab10YF: above 10-year fallow; YMI: yam–maize intercropping system; TP: teak plantation.

## 4. Discussion

The present study assessed the soil organic carbon stock and its distribution in three particle-size fractions considering five cropping systems in the Kiti sub-watershed in the Zou watershed in central Benin. This paper contributes to a growing understanding of the dynamics of soil organic carbon storage in coarse structure tropical soils in sub-Saharan Africa (SSA). The C stock in this study was estimated using the method considering the sum of the stocks of the different layers of the soil profile (0–10, 10–20, 20–30 and 30–40 cm), known as the classical method for C stock calculation. The limitation of this method is that it does not consider the variations in soil mass. To curb this, Ellert et al. [72] and Arrouays et al. [73] introduced soil equivalent masses and suggested C stocks estimation by the equivalent masses rather than the estimation by soil depth. This approach allows to reliably assess the changes in organic matter quantities linked to time or soil management practices. This method of calculation was used by Barthès et al. [14]; Aholoukpè [33] and Houssoukpèvi et al. [45].

The C stock recorded in this study ranged between 9.23 and 20.84 Mg C·ha$^{-1}$ for the layer 0–30 cm and between 11.48 and 22.20 Mg C·ha$^{-1}$ for the layer 0–40 cm. The carbon stock recorded in this study was slightly smaller than those recorded by previous studies in Benin [32,33,38,45,74]. However, the stocks recorded were higher than those recorded by Saidou et al. [42]. The low stock recorded in this study compared to previous studies could be attributed to various factors including tillage, the absence of crop residue restitution and seasonal vegetation fires which have been proven to negatively influence soil organic carbon stocks [74]. Moreover, the low stock observed in the study could also be attributed to the high proportion of concretion in the soil which can lead to a low proportion of fine particles and consequently affect soil organic carbon stocks. This is in line with research by Hairiah et al. [75] and Reichenbach et al. [76] who illustrated that the geochemical properties of the soil parent material leave a footprint that affects SOC stocks and mineral-related C stabilization mechanisms.

The soil under fallow (5YF and Ab10YF) had the highest C stock compared to teak plantation (TP) and the croplands (MCRC and YMI). The results are consistent with other work [32,77,78], highlighting that soils under fallow are enriched in organic matter from decaying litter, leaves and branches with lignified materials, which decompose progressively and replenish the soil organic carbon pool. The C stocks recorded in this study were low compared to those recorded by [75] (13.68; 12.73 and 24.40 Mg C. ha$^{-1}$ for the layer 0–20 cm) on vegetable farmland receiving organic amendment (poultry manure and sheep dung) and a 5-year fallow. These differences could be attributed to differences in farm management practices implemented in regard to vegetable farming versus staple and cash crop farming. Previous studies assessing the effect of farm management practices revealed that farm management practices, including a fallow period from five years and above, soil amendments, cover crops, mulching and crop residues restitution have a positive effect on C stocks in the soil [2,29,32,33]. The cultivated lands (MCRC and YMI) recorded the lowest C stocks. This could be attributed to the low crop residue restitution and the tillage system which can induce soil aggregate crumbling and therefore rapid carbon mineralization. The C stock observed in the soil under teak plantation was lower compared to the 30.5 and 31.4 Mg C·ha$^{-1}$ recorded by Houssoukpèvi et al. [45] for cropland and tree plantations in southern Benin. This could be attributed to the age of the plantation and the uneven exploitation scheme, making the plantation have a scattered structure and therefore a lower carbon input. Since existing studies reporting on C stocks in Benin were conducted at different layers making their comparison challenging, an extrapolation offers the possibility to compared different cropping systems on the basis of the surface layer of 0–30 cm. This extrapolation depicted that for the Acrisol in southern Benin, C stocks of 73 Mg C·ha$^{-1}$ under fallow, 41 Mg C·ha$^{-1}$ under vegetable farming systems with chicken manure, 38 Mg C·ha$^{-1}$ with sheep ruminant dung [32] and 32 Mg C·ha$^{-1}$ under the maize–mucuna cropping system [14]. This confirms that farm management practices have a substantial effect on soil organic carbon stocks. Since limited studies have focused on C stock evaluation across different farming systems, different agroecological zones, and different soil types as well as the long-term influence of these systems, more in-depth studies will help to identify and implement sustainable farming systems for better carbon sequestration and resilient food systems. In addition, as watersheds have a proven propensity to soil erosion [79], understanding the impact of landform on soil organic carbon storage could have a great contribution to developing sustainable farming systems at the watershed scale.

The particle-size distribution of organic carbon in the different particle-size fractions indicated that the non-particulate organic carbon fraction, associated with silt-clay (<53 µm), held the largest contribution to the total organic carbon. The contributions of the coarse particulate organic matter (cPOM) to the total soil organic carbon reserves were the lowest in all the cropping systems. The carbon associated with the organo-mineral fraction are localized in clay and silt bonds which protect the carbon from mineralization. Indeed, under conditions highly favorable to biological decomposition and humification, such as those in tropical regions, particulate organic matter is exposed to mineralization processes and therefore represents a small portion of the total organic carbon pool in the soil [80]. The two particulate fractions, cPOM and fPOM, have been proven to be the most affected by cropping systems [81]. These results are consistent with previous studies that emphasized the vulnerability of this fraction to mineralization processes [65,81]. These fractions, being free from soil mineral particles, are more accessible to microorganisms. The low cPOM and fPOM in the cultivated land, could be explained by tillage and soil erosion. Tillage induces soil aggregate breakdown and accelerates organic carbon mineralization. Furthermore, the cPOM and fPOM are lighter fraction and therefore susceptible to be streamed away during storm rain. For example, Akplo [79] pointed out in the Zou watershed that the particulate portion (cPOM + fPOM) of soil organic carbon was significantly affected by soil erosion.

Although the carbon content in the biomass and soil amendments applied to soil are generally accumulated more in the fPOM and cPOM fractions, the long-term accumulation of carbon in soil is predominantly determined by the carbon in the silt and clay fractions, NOM [82]. The higher concentration of the cPOM and fPOM in the two fallow lands can be attributed to the shoots and residues from the thick vegetation that accumulated during the fallow period. The fine compartment organic matter (NOM) is associated with micro-aggregates that protect organic carbon by adsorption and occlusion on mineral surfaces. The abundance of fine elements favors stabilization of soil organic carbon at a higher level of dynamic equilibrium, good structural stability of soil and makes the system more sustainable. According to our results, NOM is significantly higher in the fallow land. The biologically and chemically active fractions, fPOM and cPOM, belonging to the labile compartment is very sensitive to cultivation practices [83]. Ploughing is considered as an unfavorable factor for the storage of organic matter in the soil [84]. It favors the destruction of soil micro-aggregates and accelerates soil organic carbon mineralization. This explains the low concentrations observed in the cultivated lands. Long-term fallows remain good sustainable land management practices. However, in the current context, where croplands are shrinking in favor of urbanization and development, yet food need is growing and cropland expansion is limited, there is a need for in-depth studies establishing more sustainable and resilient intensification of farming systems [85,86].

## 5. Conclusions

In this paper, we assessed soil organic carbon stocks and its particle-size fractionation across different cropping systems. The fallow lands, Ab10YF and 5YF, exhibited the highest C stocks 17.74 and 22.20 mg C·ha$^{-1}$, while cultivated land under tillage MCRC depicted the lowest C stocks. The three organic carbon fractions showed significant variation across the cropping systems, with the NOM fraction holding the largest contributions to total soil organic carbon for all the cropping systems. The cPOM and fPOM were most influenced by the cropping systems with the highest concentration observed in Ab10YF and 5YF. The hypotheses established at the beginning of the study, stating that fallow land and teak plantations store more carbon than cultivated lands and that the POM fraction is more sensitive to cropping system characteristics than NOM is partially confirmed since the teak plantation recorded a lower C stock. The implementation of sustainable soil fertility management practices, including efficient restitution of crop residues, mulching, legume intercropping, long rotation cycles, use of dual-purpose crops and avoiding seasonal fires are necessary to improve soil organic storage in agricultural soil in the watershed. Moreover, agricultural extension officers, policy-makers and officials of the ministry in charge of agriculture have the significant role in supporting farmers in the sustainable intensification of cropping systems and the regulation and enforcement for the respect of pastoral transhumance corridors.

**Author Contributions:** Conceptualization: A.M.A.A., F.K.A. and T.M.A.; data curation: A.M.A.A. and T.M.A.; formal analysis: A.M.A.A. and F.K.A.; finding acquisition and investigation: F.K.A. and T.M.A.; methodology: A.M.A.A., F.K.A. and T.M.A.; project administration: F.K.A., H.A.A. and P.H.; supervision: F.K.A.; validation: H.A.A. and P.H.; visualization: F.K.A., T.M.A.; writing—original draft: A.M.A.A. and T.M.A.; writing—review and editing: A.M.A.A., F.K.A., T.M.A., H.A.A. and P.H. All authors have read and agreed to the published version of the manuscript.

**Funding:** This research received no external funding.

**Institutional Review Board Statement:** Not applicable.

**Informed Consent Statement:** Not applicable.

**Data Availability Statement:** Data available upon request from the corresponding author.

**Acknowledgments:** We are grateful to the International Atomic Energy Agency (IAEA), which provided the equipment used for laboratory analysis, through the Regional Project RAF5075.

**Conflicts of Interest:** The authors declare that they have no known competing financial interests or personal relationships that could appear to influence the work reported in this paper. The funders had no role in the design of the study; in the collection, analysis, or interpretation of the data; in the writing of the manuscript, or in the decision to publish the results.

## References

1. Searchinger, T.; Waite, R.; Hanson, C.; Ranganathan, J.; Dumas, P.; Matthews, E.; Klirs, C. *Creating a Sustainable Food Future: A Menu of Solutions to Feed Nearly 10 Billion People by 2050*; Final Report; WRI: Washington, DC, USA, 2019.
2. Lal, R.; Negassa, W.; Lorenz, K. Carbon sequestration in soil. *Curr. Opin. Environ. Sustain.* **2015**, *15*, 79–86. [CrossRef]
3. Fischer, R.A.; Connor, D.J. Issues for cropping and agricultural science in the next 20 years. *Field Crops Res.* **2018**, *222*, 121–142. [CrossRef]
4. Bekunda, M.; Sanginga, N.; Woomer, P.L. Chapter Four—Restoring Soil Fertility in Sub-Sahara Africa. In *Advances in Agronomy*; Sparks, D.L., Ed.; Academic Press: Cambridge, MA, USA, 2010; Volume 108, pp. 183–236.

5. Kouelo, A.; Houngnandan, P.; Azontondé, H.; Benmansour, M.; Rabesirana, N.; Mabit, L. Assessment of the level of soil degradation in three watersheds affected by intensive farming practices in Benin. *J. Exp. Biol. Agric. Sci.* **2015**, *3*, 529–540.
6. Mucheru-Muna, M.; Mugendi, D.; Pypers, P.; Mugwe, J.; Kung'U, J.; Vanlauwe, B.; Merckx, R. Enhancing Maize Productivity and Profitability Using Organic Inputs and Mineral Fertilizer in Central Kenya Small-Hold Farms. *Exp. Agric.* **2014**, *50*, 250–269. [CrossRef]
7. Blum, W.E.H. Functions of Soil for Society and the Environment. *Rev. Environ. Sci. Bio Technol.* **2005**, *4*, 75–79. [CrossRef]
8. Koch, A.; McBratney, A.; Adams, M.; Field, D.; Hill, R.; Crawford, J.; Minasny, B.; Lal, R.; Abbott, L.; O'Donnell, A.; et al. Soil Security: Solving the Global Soil Crisis. *Glob. Policy* **2013**, *4*, 434–441. [CrossRef]
9. Lal, R. Sequestration of atmospheric $CO_2$ in global carbon pools. *Energy Environ. Sci.* **2008**, *1*, 86–100. [CrossRef]
10. Bashagaluke, J.B.; Logah, V.; Opoku, A.; Sarkodie-Addo, J.; Quansah, C. Soil nutrient loss through erosion: Impact of different cropping systems and soil amendments in Ghana. *PLoS ONE* **2018**, *13*, e0208250. [CrossRef]
11. Azontonde, H.A.; Igue, A.M.; Dagbenonbakin, D.G. La Carte de Fertilité des Sols du Bénin par Zone Agro-Ecologique du Benin Rapport Final; Ministère de l'Agriculture de l'Elevage et de la Pêche Cotonou: Cotonou, Bénin, 2016; p. 120.
12. Mokwunye, A.U.; Bationo, A. Innovations as Key to the Green Revolution in Africa: Dordrecht. In *Meeting the Demands for Plant Nutrients for an African Green Revolution: The Role of Indigenous Agrominerals*; Bationo, A., Waswa, B., Okeyo, J.M., Maina, F., Kihara, J.M., Eds.; Springer: Dordrecht, The Netherlands, 2011; pp. 19–29.
13. Mkonda, M.Y.; He, X. The Influence of Soil Organic Carbon and Climate Variability on Crop Yields in Kongwa District, Tanzania. *Environ. Manag.* **2022**, 1–9. [CrossRef]
14. Barthès, B.; Azontonde, A.; Blanchart, E.; Girardin, C.; Villenave, C.; Lesaint, S.; Oliver, R.; Feller, C. Effect of a legume cover crop (Mucuna pruriens var. utilis) on soil carbon in an Ultisol under maize cultivation in southern Benin. *Soil Use Manag.* **2004**, *20*, 231–239. [CrossRef]
15. Conforti, M.; Buttafuoco, G.; Leone, A.P.; Aucelli, P.P.C.; Robustelli, G.; Scarciglia, F. Studying the relationship between water-induced soil erosion and soil organic matter using Vis–NIR spectroscopy and geomorphological analysis: A case study in southern Italy. *Catena* **2013**, *110*, 44–58. [CrossRef]
16. Hancock, G.R.; Murphy, D.; Evans, K.G. Hillslope and catchment scale soil organic carbon concentration: An assessment of the role of geomorphology and soil erosion in an undisturbed environment. *Geoderma* **2010**, *155*, 36–45. [CrossRef]
17. Baveye, P.C.; Schnee, L.S.; Boivin, P.; Laba, M.; Radulovich, R. Soil Organic Matter Research and Climate Change: Merely Re-storing Carbon Versus Restoring Soil Functions. *Front. Environ. Sci.* **2020**, *8*, 579904. [CrossRef]
18. Paul, B.K.; Vanlauwe, B.; Ayuke, F.; Gassner, A.; Hoogmoed, M.; Hurisso, T.T.; Koala, S.; Lelei, D.; Ndabamenye, T.; Six, J.; et al. Medium-term impact of tillage and residue management on soil aggregate stability, soil carbon and crop productivity. *Agric. Ecosyst. Environ.* **2013**, *164*, 14–22. [CrossRef]
19. Menta, C. Soil fauna diversity-function, soil degradation, biological indices, soil restoration. *Biodivers. Conserv. Util. A Divers. World* **2012**, 59–94.
20. Sarkar, R.; Sarkar, D.; Sinha, A.K.; Danish, S.; Bhattacharya, P.M.; Mukhopadhyay, P.; Salmen, S.H.; Ansari, M.J.; Datta, R. Soil organic carbon and labile and recalcitrant carbon fractions attributed by contrasting tillage and cropping systems in old and recent alluvial soils of subtropical eastern India. *PLoS ONE* **2021**, *16*, e0259645.
21. Blanco-Moure, N.; Gracia, R.; Bielsa, A.C.; López, M.V. Soil organic matter fractions as affected by tillage and soil texture under semiarid Mediterranean conditions. *Soil Tillage Res.* **2016**, *155*, 381–389. [CrossRef]
22. Tiefenbacher, A.; Sandén, T.; Haslmayr, H.-P.; Miloczki, J.; Wenzel, W.; Spiegel, H. Optimizing Carbon Sequestration in Croplands: A Synthesis. *Agronomy* **2021**, *11*, 882. [CrossRef]
23. Amoakwah, E.; Lucas, S.T.; Didenko, N.A.; Rahman, M.A.; Islam, K.R. Impact of deforestation and temporal land-use change on soil organic carbon storage, quality, and lability. *PLoS ONE* **2022**, *17*, e0263205. [CrossRef]
24. Gonzalez-Sanchez, E.J.; Veroz-Gonzalez, O.; Conway, G.; Moreno-Garcia, M.; Kassam, A.; Mkomwa, S.; Ordoñez-Fernandez, R.; Triviño-Tarradas, P.; Carbonell-Bojollo, R. Meta-analysis on carbon sequestration through Conservation Agriculture in Africa. *Soil Tillage Res.* **2019**, *190*, 22–30. [CrossRef]
25. Lorenz, K. Organic urban agriculture. *Soil Sci.* **2015**, *180*, 146–153. [CrossRef]
26. Saviozzi, A.; Vanni, G.; Cardelli, R. Carbon mineralization kinetics in soils under urban environment. *Appl. Soil Ecol.* **2014**, *73*, 64–69. [CrossRef]
27. Poffenbarger, H.J.; Barker, D.W.; Helmers, M.J.; Miguez, F.E.; Olk, D.C.; Sawyer, J.E.; Six, J.; Castellano, M.J. Maximum soil organic carbon storage in Midwest U.S. cropping systems when crops are optimally nitrogen-fertilized. *PLoS ONE* **2017**, *12*, e0172293. [CrossRef]
28. Croft, H.; Kuhn, N.J.; Anderson, K. On the use of remote sensing techniques for monitoring spatio-temporal soil organic carbon dynamics in agricultural systems. *CATENA* **2012**, *94*, 64–74. [CrossRef]
29. Luo, Z.; Feng, W.; Luo, Y.; Baldock, J.; Wang, E. Soil organic carbon dynamics jointly controlled by climate, carbon inputs, soil properties and soil carbon fractions. *Glob. Chang. Biol.* **2017**, *23*, 4430–4439. [CrossRef]
30. Mugwe, J.; Ngetich, F.; Otieno, E.O. Integrated Soil Fertility Management in sub-Saharan Africa: Evolving Paradigms toward Integration. In *Zero Hunger. Encyclopedia of the UN Sustainable Development Goals*; Leal Filho, W., Azul, A.M., Brandli, L., Özuyar, P.G., Wall, T., Eds.; Springer International Publishing: Cham, Germany, 2019.

31. Hobley, E.; Wilson, B.; Wilkie, A.; Gray, J.; Koen, T. Drivers of soil organic carbon storage and vertical distribution in Eastern Australia. *Plant Soil* **2015**, *390*, 111–127. [CrossRef]
32. Koussihouèdé, K.I.H.; Aholoukpè, H.N.S.; Assogba, K.F.V.; Amadji, G.L. Soil organic carbon status in a vegetable cropping system in Southern Benin: A rapid assessment. *Afr. J. Soil Sci.* **2017**, *5*, 410–419.
33. Aholoukpè, H. Matière Organique du sol et Développement du Palmier à Huile sous Différents Modes de Gestion des Feuilles d'élagage: Cas des Palmeraies Villageoises du Département du Plateau au Bénin. Ph.D. Thesis, Université d'Abomey-Calavi, Montpellier, France, 2013.
34. Carter, M. Analysis of Soil Organic Matter Storage in Agroecosystems. In *Structure and Organic Matter Storage in Agricultural Soils*; CRC Press: Boca Raton, FL, USA, 2020; pp. 3–11.
35. Xu, J.; Sun, Y.; Gao, L.; Cui, X. A review of the factors influencing soil organic carbon stability. *Zhongguo Shengtai Nongye Xuebao Chin. J. Eco Agric.* **2018**, *26*, 222–230.
36. Marriott, E.E.; Wander, M.M. Total and labile soil organic matter in organic and conventional farming systems. *Soil Sci. Soc. Am. J.* **2006**, *70*, 950–959. [CrossRef]
37. Ghimire, R.; Lamichhane, S.; Acharya, B.S.; Bista, P.; Sainju, U.M. Tillage, crop residue, and nutrient management effects on soil organic carbon in rice-based cropping systems: A review. *J. Integr. Agric.* **2017**, *16*, 1–15. [CrossRef]
38. Yang, X.; Meng, J.; Lan, Y.; Chen, W.; Yang, T.; Yuan, J.; Liu, S.; Han, J. Effects of maize stover and its biochar on soil $CO_2$ emissions and labile organic carbon fractions in Northeast China. *Agric. Ecosyst. Environ.* **2017**, *240*, 24–31. [CrossRef]
39. Choudhury, S.G.; Srivastava, S.; Singh, R.; Chaudhari, S.K.; Sharma, D.K.; Singh, S.K.; Sarkar, D. Tillage and residue management effects on soil aggregation, organic carbon dynamics and yield attribute in rice–wheat cropping system under reclaimed sodic soil. *Soil Tillage Res.* **2014**, *136*, 76–83. [CrossRef]
40. Yoo, G.; Wander, M.M. Tillage effects on aggregate turnover and sequestration of particulate and humified soil organic carbon. *Soil Sci. Soc. Am. J.* **2008**, *72*, 670–676. [CrossRef]
41. Semenov, V.M.; Lebedeva, T.N.; Pautova, N.B. Particulate Organic Matter in Noncultivated and Arable Soils. *Eurasian Soil Sci.* **2019**, *52*, 396–404. [CrossRef]
42. Saidou, A.; Balogoun, I.; Kone, B.; Gnangle, C.; Aho, N. Effet d'un système agroforestier à karité (Vitellaria paradoxa cf gaertn) sur le sol et le potentiel de production du maïs (Zea maize) en zone Soudanienne du Bénin. *Int. J. Biol. Chem. Sci.* **2012**, *6*, 2066–2082.
43. Bello, O.; Saïdou, A.; Ahoton, E.; Avaligbé, J.; Ezin, A.; Akponikpè, P.; Aho, N. Assessment of organic carbon stock in cashew plantations (*Anacardium occidentale* L.) in Benin (West Africa). *Int. J. Agric. Environ. Res.* **2017**, *3*, 3471–3495.
44. Azontonde, H.A. Dynamique de la Matiere Organique et de l'azote dans le France Mucuna-Mais sur un sol Ferrallitique (Terres de Barre) au Sud-Benin. Ph.D. Thesis, des Scientifiques de l'IRD, Montpellier, France, 2000.
45. Houssoukpèvi, I.A.; Aholoukpè, H.N.S.; Fassinou, D.J.M.; Rakotondrazafy, M.N.; Amadji, G.L.; Chapuis-Lardy, L.; Chevallier, T. Biomass and soil carbon stocks of the main land use of the Allada Plateau (Southern Benin). *Carbon Manag.* **2022**, *13*, 249–265. [CrossRef]
46. Djeui, N.; de Boisseon, P.; Gavinelli, E. Statut organique d'un sol ferrallitique du Sud-Bénin sous forêt et différents systèmes de cultures. *Cah. ORSTOM Série Pédologie* **1992**, *27*, 5–22.
47. Giller, K.E.; Tittonell, P.; Rufino, M.C.; van Wijk, M.T.; Zingore, S.; Mapfumo, P.; Adjei-Nsiah, S.; Herrero, M.; Chikowo, R.; Corbeels, M.; et al. Communicating complexity: Integrated assessment of trade-offs concerning soil fertility management within African farming systems to support innovation and development. *Agric. Syst.* **2011**, *104*, 191–203. [CrossRef]
48. Roudier, P.; Sultan, B.; Quirion, P.; Berg, A. The impact of future climate change on West African crop yields: What does the recent literature say? *Glob. Environ. Chang.* **2011**, *21*, 1073–1083. [CrossRef]
49. Atchada, C.C.; Zoffoun, A.G.; Akplo, T.M.; Azontonde, A.H.; Tente, A.B.; Djego, J.G. Modes d'utilisation des terres et stock de carbone organique du sol dans le bassin supérieur de Magou au Bénin. *Int. J. Biol. Chem. Sci.* **2018**, *12*, 2818–2829. [CrossRef]
50. CPCS. Classification des sols. Labo. de Géologie Pédologie, Grignon, Ronéo. Labo. de Géologie Pédologie, Grignon, Ronéo 1967, (Commissiom de pédologie et de cartographie des sols), 87p. Available online: https://horizon.documentation.ird.fr/exl-doc/pleins_textes/divers16-03/12186.pdf (accessed on 25 September 2022).
51. FAO. *World Reference Base for Soil Resources 2014*; FAO: Rome, Italy, 2015.
52. Youssouf, I.; Lawani, M. *Les Sols B Ninois: Classification Dans la Référence Mondiale, Quatorzieme Reunion du sous-Comite Ouest et Centre Africain de Correlation des Sols Pour la Mise en Valeur des Terre*; FAO: Rome, Italy, 2016.
53. Maliki, R.; Toukourou, M.; Sinsin, B.; Vernier, P. Productivity of Yam-Based Systems with Herbaceous Legumes and Short Fallows in the Guinea-Sudan Transition Zone of Benin. *Nutr. Cycl. Agroecosystems* **2012**, *92*, 9–19. [CrossRef]
54. Eeswaran, R.; Nejadhashemi, A.P.; Faye, A.; Min, D.; Prasad, P.V.V.; Ciampitti, I.A. Current and Future Challenges and Opportunities for Livestock Farming in West Africa: Perspectives from the Case of Senegal. *Agronomy* **2022**, *12*, 1818. [CrossRef]
55. Houessou, S.O.; Dossa, L.H.; Diogo, R.V.C.; Houinato, M.; Buerkert, A.; Schlecht, E. Change and continuity in traditional cattle farming systems of West African Coast countries: A case study from Benin. *Agric. Syst.* **2019**, *168*, 112–122. [CrossRef]
56. Andrieu, N.; Vayssières, J.; Corbeels, M.; Blanchard, M.; Vall, E.; Tittonell, P. From farm scale synergies to village scale trade-offs: Cereal crop residues use in an agro-pastoral system of the Sudanian zone of Burkina Faso. *Agric. Syst.* **2015**, *134*, 84–96. [CrossRef]
57. Junge, B.; Alabi, T.; Sonder, K.; Marcus, S.; Abaidoo, R.; Chikoye, D.; Stahr, K. Use of remote sensing and GIS for improved natural resources management: Case study from different agroecological zones of West Africa. *Int. J. Remote Sens.* **2010**, *31*, 6115–6141. [CrossRef]

58. Aoudji, A.K.N.; Adégbidi, A.; Agbo, V.; Atindogbé, G.; Toyi, M.S.S.; Yêvidé, A.S.I.; Ganglo, J.C.; Lebailly, P. Functioning of farm-grown timber value chains: Lessons from the smallholder-produced teak (*Tectona grandis* L.f.) poles value chain in Southern Benin. *For. Policy Econ.* **2012**, *15*, 98–107. [CrossRef]
59. Ellison, J.; Brinkmann, K.; Diogo, R.V.C.; Buerkert, A. Land cover transitions and effects of transhumance on available forage biomass of rangelands in Benin. *Environ. Dev. Sustain.* **2022**, *24*, 12276–12310. [CrossRef]
60. Takacs, S.; Schulte to Bühne, H.; Pettorelli, N. What shapes fire size and spread in African savannahs? *Remote Sens. Ecol. Conserv.* **2021**, *7*, 610–620. [CrossRef]
61. Wollenhaupt, N.; Wolkowski, R.; Clayton, M. Mapping soil test phosphorus and potassium for variable-rate fertilizer application. *J. Prod. Agric.* **1994**, *7*, 441–448. [CrossRef]
62. Okalebo, J.R.; Gathua, K.W.; Woomer, P.L. *Laboratory Methods of Soil and Plant Analysis: A Working Manual*, 2nd ed.; Sacred Africa: Wolverhampton, UK, 2002; 26p.
63. Robinson, G.W. A new method for the mechanical analysis of soils and other dispersions. *J. Agric. Sci.* **1922**, *12*, 306–321. [CrossRef]
64. Feller, C. Organic inputs, soil organic matter and functional soil organic compartments in low-activity clay soils in tropical zones. *Soil Org. Matter Dyn. Sustain. Trop. Agric.* **1993**, 77–88.
65. Sainepo, B.M.; Gachene, C.K.; Karuma, A. Assessment of soil organic carbon fractions and carbon management index under different land use types in Olesharo Catchment, Narok County, Kenya. *Carbon Balance Manag.* **2018**, *13*, 4. [CrossRef] [PubMed]
66. Gura, I.; Mnkeni, P.; du Preez, C.; Barnard, J. Response of soil carbon fractions in a Haplic Cambisol to crop rotation systems and residue management practices under no tillage in the Eastern Cape, South Africa. *S. Afr. J. Plant Soil* **2021**, *38*, 372–380. [CrossRef]
67. Cambardella, C.A.; Elliott, E. Particulate soil organic-matter changes across a grassland cultivation sequence. *Soil Sci. Soc. Am. J.* **1992**, *56*, 777–783. [CrossRef]
68. Nelson, D.W.; Sommers, L.E. Total carbon, organic carbon, and organic matter. *Methods Soil Anal. Part 3 Chem. Methods* **1996**, *5*, 961–1010.
69. Christensen, B. Physical fractionation of soil and structural and functional complexity in organic matter turnover. *Eur. J. Soil Sci.* **2001**, *52*, 345–353. [CrossRef]
70. Shapiro, S.S.; Wilk, M.B. An analysis of variance test for normality (complete samples). *Biometrika* **1965**, *52*, 591–611. [CrossRef]
71. Bartlett, M.S. Properties of Sufficiency and Statistical Tests. *Proc. R. Soc. London Ser. A Math. Phys. Sci.* **1937**, *160*, 268–282.
72. Ellert, B.H.; Janzen, H.H.; Entz, T. Assessment of a method to measure temporal change in soil carbon storage. *Soil Sci. Soc. Am. J.* **2002**, *66*, 1687–1695. [CrossRef]
73. Arrouys, D.; Jolivet, C.; Boulonne, L.; Bodineau, G.; Ratié, C.; Saby, N.; Grolleau, E. Le réseau de Mesures de la Qualité des SFrance France (RMQS). *Etude Gest. Des Sols* **2003**, *10*, 241–250.
74. Volkoff, B.; Faure, P.; Dubroeucq, D.; Viennot, M. Estimation des stocks de carbone des sols du Bénin. *Etude Gest. Des Sols* **1999**, *6*, 115–130.
75. Hairiah, K.; van Noordwijk, M.; Sari, R.R.; Saputra, D.D.; Widianto; Suprayogo, D.; Kurniawan, S.; Prayogo, C.; Gusli, S. Soil carbon stocks in Indonesian (agro) forest transitions: Compaction conceals lower carbon concentrations in standard accounting. *Agric. Ecosyst. Environ.* **2020**, *294*, 106879. [CrossRef]
76. Reichenbach, M.; Fiener, P.; Garland, G.; Griepentrog, M.; Six, J.; Doetterl, S. The role of geochemistry in organic carbon stabilization against microbial decomposition in tropical rainforest soils. *Soil* **2021**, *7*, 453–475. [CrossRef]
77. Parras-Alcántara, L.; Díaz-Jaimes, L.; Lozano-García, B. Management effects on soil organic carbon stock in Mediterranean open rangelands—Treeless grasslands. *Land Degrad. Dev.* **2015**, *26*, 22–34. [CrossRef]
78. Sarkar, D.; Bungbungcha Meitei, C.; Baishya, L.K.; Das, A.; Ghosh, S.; Chongloi, K.L.; Rajkhowa, D. Potential of fallow chronosequence in shifting cultivation to conserve soil organic carbon in northeast India. *CATENA* **2015**, *135*, 321–327. [CrossRef]
79. Akplo, T.M.; Alladassi, F.K.; Houngnandan, P.; Saidou, A.; Benmansour, M.; Azontonde, H.A. Mapping the risk of soil erosion using RUSLE, GIS and remote sensing: A case study of Zou watershed in central Benin. *Moroc. J. Agric. Sci.* **2020**, 281–290.
80. Zanatta, J.A.; Bayer, C.; Dieckow, J.; Vieira, F.C.B.; Mielniczuk, J. Soil organic carbon accumulation and carbon costs related to tillage, cropping systems and nitrogen fertilization in a subtropical Acrisol. *Soil Tillage Res.* **2007**, *94*, 510–519. [CrossRef]
81. Yu, P.; Li, Q.; Jia, H.; Zheng, W.; Wang, M.; Zhou, D. Carbon stocks and storage potential as affected by vegetation in the Songnen grassland of northeast China. *Quat. Int.* **2013**, *306*, 114–120. [CrossRef]
82. Chivenge, P.P.; Murwira, H.K.; Giller, K.E.; Mapfumo, P.; Six, J. Long-term impact of reduced tillage and residue management on soil carbon stabilization: Implications for conservation agriculture on contrasting soils. *Soil Tillage Res.* **2007**, *94*, 328–337. [CrossRef]
83. Kirchmann, H.; Haberhauer, G.; Kandeler, E.; Sessitsch, A.; Gerzabek, M.H. Effects of level and quality of organic matter input on carbon storage and biological activity in soil: Synthesis of a long-term experiment. *Glob. Biogeochem. Cycles* **2004**, *18*, 1–9. [CrossRef]
84. Ouédraogo, E.; Mando, A.; Stroosnijder, L. Effects of tillage, organic resources ertilizeren fertiliser on soil carbon dynamics and crop nitrogen uptake in semi-arid West Africa. *Soil Tillage Res.* **2006**, *91*, 57–67. [CrossRef]

85. Mehra, P.; Baker, J.; Sojka, R.E.; Bolan, N.; Desbiolles, J.; Kirkham, M.B.; Ross, C.; Gupta, R. Chapter Five—A Review of Tillage Practices and Their Potential to Impact the Soil Carbon Dynamics. In *Advances in Agronomy*; Sparks, D.L., Ed.; Academic Press: Cambridge, MA, USA, 2018; Volume 150, pp. 185–230.
86. Garnett, T.; Appleby, M.C.; Balmford, A.; Bateman, I.J.; Benton, T.G.; Bloomer, P.; Burlingame, B.; Dawkins, M.; Dolan, L.; Fraser, D.; et al. Sustainable Intensification in Agriculture: Premises and Policies. *Science* **2013**, *341*, 33–34. [CrossRef] [PubMed]

*Article*

# Biomass-Derived N-Doped Activated Carbon from Eucalyptus Leaves as an Efficient Supercapacitor Electrode Material

Dinesh Bejjanki [1], Praveen Banothu [1], Vijay Bhooshan Kumar [2,*] and Puttapati Sampath Kumar [1,*]

1. Department of Chemical Engineering, National Institute of Technology Warangal, Hanamkonda 506004, Telangana, India
2. The Shmunis School of Biomedicine and Cancer Research, George S. Wise Faculty of Life Sciences, Tel Aviv University, Tel Aviv 6997801, Israel
* Correspondence: vijaybooshan@mail.tau.ac.il or vijaybhushan86@gmail.com (V.B.K.); pskr@nitw.ac.in (P.S.K.)

**Abstract:** Biomass-derived activated carbon is one of the promising electrode materials in supercapacitor applications. In this work bio-waste (oil extracted from eucalyptus leaves) was used as a carbon precursor to synthesize carbon material with $ZnCl_2$ as a chemical activating agent and activated carbon was synthesized at various temperatures ranging from 400 to 800 °C. The activated carbon at 700 °C showed a surface area of 1027 $m^2\ g^{-1}$ and a specific capacitance of 196 $F\ g^{-1}$. In order to enhance the performance, activated carbon was doped with nitrogen-rich urea at a temperature of 700 °C. The obtained activated carbon and N-doped activated carbon was characterized by phase and crystal structural using (XRD and Raman), morphological using (SEM), and compositional analysis using (FTIR). The electrochemical measurements of carbon samples were evaluated using an electrochemical instrument and NAC-700 °C exhibited a specific capacitance of 258 $F\ g^{-1}$ at a scan rate of 5 $mV\ s^{-1}$ with a surface area of 1042 $m^2\ g^{-1}$. Thus, surface area and functionalizing the groups with nitrogen showed better performance and it can be used as an electrode material for supercapacitor cell applications.

**Keywords:** eucalyptus leaves; activated carbon; nitrogen doping; chemical activation; supercapacitor

---

**Citation:** Bejjanki, D.; Banothu, P.; Kumar, V.B.; Kumar, P.S. Biomass-Derived N-Doped Activated Carbon from Eucalyptus Leaves as an Efficient Supercapacitor Electrode Material. *C* **2023**, *9*, 24. https://doi.org/10.3390/c9010024

Academic Editor: Enrico Andreoli

Received: 20 December 2022
Revised: 19 January 2023
Accepted: 23 January 2023
Published: 17 February 2023

**Copyright:** © 2023 by the authors. Licensee MDPI, Basel, Switzerland. This article is an open access article distributed under the terms and conditions of the Creative Commons Attribution (CC BY) license (https://creativecommons.org/licenses/by/4.0/).

## 1. Introduction

Energy is necessary in day today life. In future, energy generation with sufficient and sustainable methods will be a major concern. Currently, over 75% of the energy used by society is produced by fossil fuels, including coal, natural gas, and oil, neither of which are renewable. As a result, energy must always be replaced by renewable energy before fossil fuels run out. Excessive use of fossil fuels leads to global warming. It releases harmful gases directly into the environment [1,2]. To address the challenges related to environmental pollution and energy production, fuels with petroleum must be transformed to renewable and sustainable energy sources. Unutilized waste materials and biomass can be used effectively in synthesizing of carbon materials for energy storage and conversion devices. The carbonization process carried out by heating biomasses under inert gas condition and high temperature. In contrast, the heteroatoms (oxygen, sluphur and nitrogen etc.,) in the networks of the biological macromolecules escapes there by leaving the carbon skeletons with the porous shape. So when residual carbon skeletons are activated, they may form linked 3D structures with considerably high conductivity, porosity and surface area, making them attractive candidates in energy storage applications. There are various biomass-derived carbon materials using highly effective green energy storage devices. Finding alternative energy resources that use green energy (biomass) and clean energy is essential [3,4]. So far, several storage technologies have been developed such as capacitors, batteries, fuel cells, and supercapacitors [1,2,5]. Supercapacitors are in the field of electrochemical devices with their remarkable fast charging speed, high power density, light weight, safe operation, and long life cycle [3,4,6]. Supercapacitors are divided into symmetric and asymmetric supercapacitors [5,7]. Electrochemical double-layer capacitors (EDLC) and pseudo capacitors are classified as symmetric supercapacitors, while the hybrid capacitor is an asymmetric supercapacitor. The charge storage mechanism classifies the behavior of supercapacitor. The EDLC stores energy electrostatically while pseudo capacitors store energy via electrochemical redox or faradaic reactions. The Supercapacitor consists of 4 major components they are electrolyte, electrode, current collector, and separator. Among all the components, the performance of supercapacitor is mainly based on the electrode material [8]. Activated carbon, carbon nanotubes, and carbon nanofibers are utilized as electrode materials in EDLC, whereas metal-based oxides ($RuO_2$, $Co_3O_4$, $MnO_2$, NiO) and conducting polymers (polyaniline, poly-(3,4-ethylenedioxythiophene), and polypyrrole) are employed as pseudo capacitor materials, whereas the pseudocapacitors are chemical stability, high conductivity, corrosion resistance, controlled pore structure, temperature stability, environmental friendliness, and low cost [9].

In recent years, biomass-derived carbon material as electrode material gained a wide variety of attention, because of its unique properties such as high specific surface area (SSA), excellent electrical conductivity, and low production cost. Carbon materials exhibited in different morphologies, such as graphene, carbon nanotubes (CNT), carbon sphere, and carbon nanoparticle, thus have been looked into supercapacitor applications. Although, these electrode materials show short-time durability, low energy density, and high power density with less porosity [4,10,11]. The biomass-derived activated carbon has been widely used as electrode material for supercapacitor application. Considering the importance of activated carbon from biomass in an energy storage device, the literature has been carried out to know the chemical composition of various biomass such as Noeli et al. have studied the physical characteristics of dried bananas leaving and it has a carbon content of 43.5 wt.%. [12], D. Pujol et al. reported the elemental analysis of coffee waste extracted from soluble coffee industry with a carbon wt.% has 57 [13], Salwan et al. studied the chemical composition of black tea waste and algae both biomass waste has a carbon wt. percent of 30 & 28 respectively [14], and Yang Liu et al. studied biomass waste of willow leave, and which has 45 wt.% of carbon [15], Nannan et al. and Saad A et al. studied waste biomass, i.e., tremella also known as white fungus, and Egyptin mango leaves both biomass have the carbon as wt.% 40.25 and 40.7, respectively [16,17], and Grima et al., studied the carbonaceous residue of eucalyptus leaves biomass, and it has 74.5 wt.% carbon [18]. Compared to all other biomasses, the eucalyptus has the highest carbon content (approximately 75%) [18]. The chemical activation technique used to synthesize activated carbon will help to lower its issues and can increase its active sites. The chemical activating agent are $ZnCl_2$, KOH, $H_3PO_4$, NaOH, $K_2CO_3$, $H_2SO_4$, NaCl, and $CaCl_2$ [19–21]. The $ZnCl_2$ has been used as activating agent to prepare activated carbon mostly for lignocellulosic biomass due which acts as dehydrating and dampening agent during the chemical activation. The $ZnCl_2$ activation causes swelling in the cellulose structure due to electrolytic action which also leads to increasing the surface area of activated carbon. The $ZnCl_2$ activated carbon (AC) has been widely used as electrode material in energy storage application. Furthermore, AC having low graphitization degrees usually have poor electric conductivity, which significantly limits the quick charge and discharge, especially at higher current densities. It's been proven that modifying AC with heteroatom species such as (oxygen, sulphur, and nitrogen functional groups), not only alter the conductivity of carbon network, but also helps in the permeability ions into electrode and electrolytes.

In this study, the chemical activation technique has been employed to synthesize activated carbon, using the carbonation method at various temperatures from 400 °C to 800 °C using $ZnCl_2$ as an activating agent. As prepared activated carbon was studied to determine the effect of temperature on its morphology and specific capacitance. Among all the AC, AC-700 °C has shown the most promising results in terms of phase, morphology, structure, composition, and electrochemical measurements. In order to synthesize N-doped activated carbon, urea was used as a nitrogen precursor and a chemical activation process was applied.

## 2. Materials and Methods

### 2.1. Materials

Raw materials required to synthesis activated carbon are eucalyptus leaves, zinc chloride ($ZnCl_2$), hydrochloric acid 35% (HCl), ethanol (>99% purity), distilled water (DI), and urea ($CH_4N_2O$) reagents were purchased from Merck India Pvt Ltd. (Singapore). Nickel foam has current collector with (0.5mm × 20cm × 30 cm) dimensions purchased from (Global nanotech Pvt. Ltd., Gujarat, India).

### 2.2. Synthesis of Activated Carbon

Eucalyptus leaves were collected from the surrounding of NIT Warangal, India. Eucalyptus leaves were pre-treated with DI water and dried, oil was extracted from the eucalyptus leaves using steam distillation. After extracting oil, the left over bio-waste (eucalyptus residue) was dried at 80 °C for 8 h. The dried biomass was crushed in fine powder by using a mortar and pestle and sieved with a mesh size of 60 to 80. The resultant eucalyptus leaves powder (ELP) was carbonized in tubular furnace under an inert atmosphere (nitrogen) with an inlet flow rate of 75 mL min$^{-1}$. The furnace consist of PID temperature controller with long tube of length 160 cm and the diameter of 40 mm. Inside the tubular furnace a quartz boat is used to hold the sample. The ELP was carbonized at 300 °C for 1 h. @ 5 °C min$^{-1}$. Next, the samples were mixed with $ZnCl_2$ in the proportion of 1:4 ($ZnCl_2$: sample). The obtained mixture was agitated to create a homogenous slurry and dried for 12 h and at 100 °C. The resultant products were then carbonized to 400–800 °C for 1 h. @ 5 °C min$^{-1}$ under an inert atmosphere. The product was dried at 80 °C for 10 h. After the process was completed, the material was cooled and taken from the tubular furnace. The carbonized material was then rinsed with 1 M HCl till the pH reached 7 and then filtered and dried for 12 h at 100 °C. The resultant product of this post-treatment was activated carbon, abbreviated as AC. The AC was labelled as AC-x, where x ($x$ = 400 to 800 °C) represents the carbonization temperature and stored AC-x materials in sample bottles.

### 2.3. Synthesis of N-Doped Activated Carbon

As obtained eucalyptus leaves powder (ELP), 10 g was weighed and carbonized at 300 °C @ 5 °C min$^{-1}$ under inert atmosphere. As obtained sample from the previous step was mixed with $ZnCl_2$ and urea in the proportion of 1:4:2 ($ZnCl_2$: sample: urea). Next, the product was then carbonized to 700 °C @ 5 °C min$^{-1}$ for 1 h. The resultant product was washed with 1 M HCl followed by DI water until the pH attained to 7 and finally the product was dried and collected. The Eucalyptus leaves N-doped carbon materials (ELNAC) so the resultant was labelled as NAC-x, where x ($x$ = 700 °C) represents the carbonization temperature.

### 2.4. Charaterization and Electrochemical Performance Evaluations

The crystal structure of AC and NAC composition was determined using a X-ray diffraction pattern (XRD PANalytical, X'Pert-PRO MPD using Cu Kα radiation) at a wavelength of 0.15406 nm with an angle θ ranging from 10° to 70°. The scanning step size was 0.02°, and also the rate of scanning was 0.1 s per step. Raman spectroscopy (Laser Raman Spectrometer, China). Raman spectra were recorded using 523 nm laser source, at wavelength ranging from 500 to 2700 cm$^{-1}$. The different functional groups and vibrational modes associated with the (AC and NAC) material were studied using FTIR analysis in the range of 400 to 4000 cm$^{-1}$. The SEM and Raman spectroscopy analysis were carried out to understand the morphology of the material and to know to the presents of defects in the (AC and NAC).

The electrochemical performance was evaluate using three-electrode system CH instrument (model-CHI660D) in aqueous 2M KOH electrolyte. In this configuration, carbon electrodes, platinum (Pt), and Ag/AgCl, were employed as the working, counter, and reference electrodes, respectively. The working electrode concocted by means of active material (AC's and NAC), conducting graphite, and polyvinylidene fluoride (PVDF) were mixed at a weight percentage of 80:10:10, and N-Methyl-2-pyrrolidone used as a solvent to make a homogeneous slurry. After that, the obtained slurry was coated on a 1*2 cm² nickel foam and then dried at 85 °C for 4 h. As prepared working electrode used to evaluate the performance curve of cyclic voltammetry (CV), and galvanostatic charge-discharge (GCD) at different scan rates and current densities in the potential range of −1.0 to 0 V. The properties of the electrode such as specific capacitance, energy density, and power density can be calculated from the GCD curves using the below Equations (1)–(3), respectively.

$$C = \frac{I \Delta t}{m \Delta V} \quad (1)$$

$$E = \frac{C \Delta V 2}{2} \quad (2)$$

$$P = \frac{E}{\Delta t} \quad (3)$$

where, $C$ = specific capacitance (F g$^{-1}$), $\Delta t$ = discharge time difference (s), $I$ = current (mA), $m$ = mass of active material in working electrode (mg), $\Delta V$ = the discharge voltage window (V), $E$ = energy density (Wh kg$^{-1}$), P = power density (W kg$^{-1}$).

## 3. Results
### 3.1. X-ray Diffraction Spectroscopy

The XRD pattern of AC at various temperatures (400 °C to 800 °C) and NAC at 700 °C was characterized as seen in Figure 1a,b. The XRD pattern of AC and NAC exhibits two diffraction peaks at 2θ values of 24° and 44° which correlate to the miller indices of (002) and (100), respectively. The obtained peak was broad shows that an amorphous carbon structure with a low degree of graphitization [22,23]. This result shows that as we increase the carbonization temperature leads to the carbon material becomes more graphitic and increases the conductivity of a sample [24].

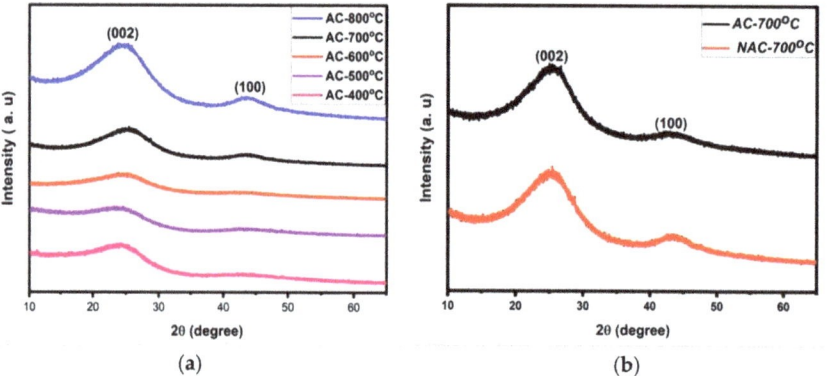

**Figure 1.** XRD pattern of (**a**) various temperature AC (**b**) AC-700 °C and NAC-700 °C.

Table 1 Represents the interlayer d-spacing and dimensions of microcrystalline of various temperature carbon materials. Corresponding to Table 1, the increase in activation temperature guides to a change in the stack height ($L_c$) and microcrystalline structure. The stack height ($L_c$) is related to the specific surface area provided by empirical formula (SSA$_{xrd}$ = 2/$\rho_{xrd}$ $L_c$), whereas $\rho_{xrd}$ is calculated from $\rho_{xrd}$= (d$_{002(graphite)}$/d$_{002}$) × $\rho_{graphite}$ and ($L_a$) is stack width. Based on the SSA$_{xrd}$ formula, the stack height ($L_c$) and surface area are inversely proportional. If the $L_c$ value is low, it means that the surface area from XRD is high and vice versa. From Table 1, AC-700 has the least stack height ($L_c$) of 9.39 °A, which shows that AC-700 has the highest specific surface area compared to various temperature electrodes [25,26].

**Table 1.** Interlayer d-spacing and microcrystalline dimensions of carbon material at various temperatures. (* is corresponding to the XRD major peak).

| Sample Code | 2θ | | Inter Layer (nm) | | Micro Crystalline Dimension | | $L_c/L_a$ | Np | $SSA_{xrd}$ |
|---|---|---|---|---|---|---|---|---|---|
| | * C002 | * C100 | d002 | d100 | $L_c$ | $L_a$ | | | |
| AC-800 | 24.35 | 43.72 | 0.365 | 0.2067 | 0.957 | 3.506 | 0.273 | 2.621 | 1008.34 |
| AC-700 | 24.85 | 43.87 | 0.357 | 0.2061 | 0.939 | 3.006 | 0.312 | 2.625 | 1027.74 |
| AC-600 | 24.26 | 43.71 | 0.366 | 0.2068 | 0.989 | 2.626 | 0.376 | 2.699 | 975.73 |
| AC-500 | 23.92 | 43.96 | 0.371 | 0.2072 | 1.000 | 2.156 | 0.463 | 2.692 | 965.10 |
| AC-400 | 23.85 | 43.01 | 0.372 | 0.2100 | 1.024 | 2.854 | 0.358 | 2.748 | 942.54 |
| NAC-700 | 25.05 | 43.49 | 0.354 | 0.2078 | 0.925 | 2.917 | 0.317 | 2.607 | 1042.92 |

*3.2. Raman Spectroscopy*

The Raman spectroscopy technique is commonly used to characterize carbon materials due to its ability to reveal the disordered structure of carbon molecules. The Raman spectrum of AC-700 °C and NAC-700 °C exhibits two peaks that corresponds to the D-band at (1360 cm$^{-1}$) and G-band at (1585 cm$^{-1}$) while the D-band represent the disorder in structure, and G-band represents vibration of carbon atoms, correspondingly as seen in (Figure 2). The intensity ratio of the D and G band is utilized to assess the level of graphitization in AC and NAC [23]. The intensity ratio ($I_D/I_G$) of AC and NAC are 0.9995 and 1.05, respectively. It indicates that the effect of N-doping guides the degree of the disorder and incorporation of heterogeneous atom (nitrogen) into the graphite layer therefore it can relate to, higher the intensity ratio the higher is the degree of disorder [27].

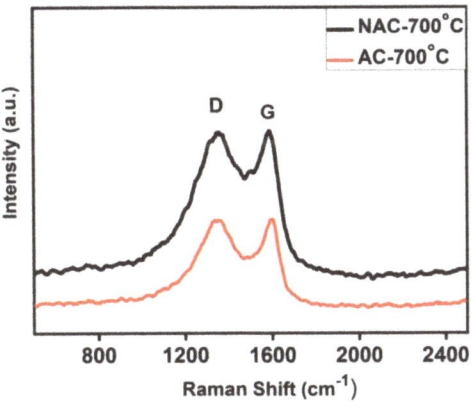

**Figure 2.** Raman spectroscopy of AC-700 °C and NAC-700 °C.

*3.3. Scanning Electronic Microscopy (SEM)*

The morphological structures and characteristics of untreated, AC (at different carbonization temperatures), and NAC were analyzed by using the SEM technique, and Figure 3b–f shows the microstructure of AC samples of temperature from 400 °C to 800 °C at different magnification. The observation of micrographs (SEM Image) having good porosity with macro pores in nature has resulted from the synthesized AC. As can be seen from Figure 3a, there are some vacant sites after pre-carbonization (untreated). Once it is activated with chemical activating agent guides to activate the inactive sites and as increasing the carbonization temperature leads to more active site formation so that the porosity of AC has been increased which corresponds to increasing the specific surface area of AC [26]. As observed from Figure 3e,g, most of the inactive site are active as carbonization temperature reaches to 700 °C. From Figure 3f, as the temperature increase further lead to damaging pores as well as non-uniform porous material. Further, the chemical composition of AC at different carbonization temperatures and NAC was determined using energy dispersive spectroscopy (EDS). A summary of the chemical composition of the different samples is presented in Figure S1 and Table S1. According to Table S1, the weight and atomic percentage of carbon increase as the carbonization temperature increases. Low to high heating rates increased the carbon content of the AC from 80.2 to 97.5 wt.%, whereas low to high heating rates decreased the oxygen content from 15.0 to 0%. Therefore, the formation of porosities in AC facilitates the mobility of ions between both the electrode and electrolyte during the charge and discharge cycle [28].

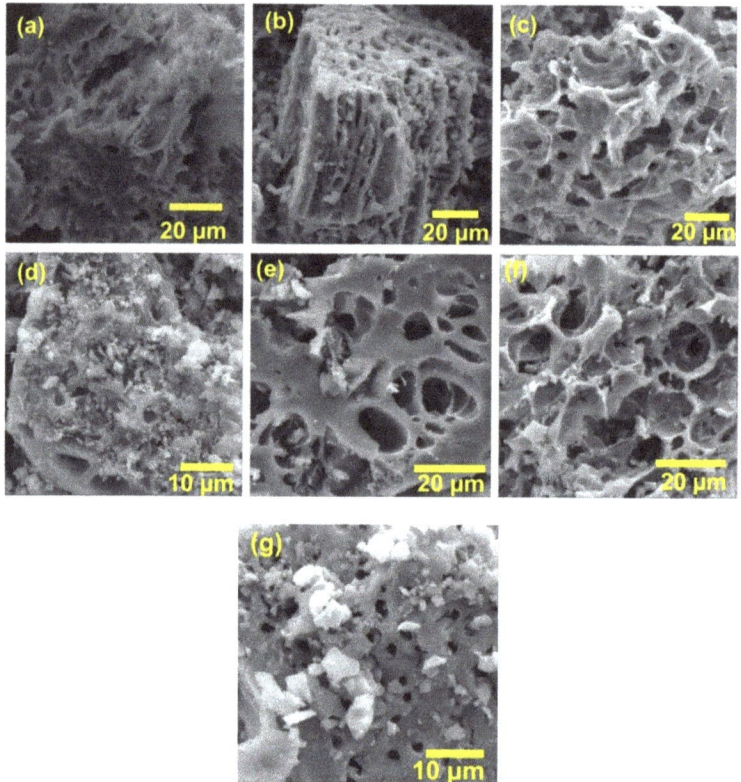

**Figure 3.** SEM images of (**a**) untreated, (**b**) AC-400 °C, (**c**) AC-500 °C, (**d**) AC-600 °C, (**e**) AC-700 °C, (**f**) AC-800 °C, and (**g**) NAC-700 °C.

*3.4. Fourier-Transform Infrared Spectroscopy (FTIR)*

The FTIR analysis of AC at a carbonization temperature of 700 °C was analyzed using the FTIR spectrometer within the scale of 400–4000 cm$^{-1}$. Figure 4a shows The sharp band around 3650 cm$^{-1}$ was observed and represents the alcohol groups of O-H stretching vibration [29]. The small band were appearing at 3135 cm$^{-1}$ and 2895 cm$^{-1}$ correspond to the alcohol groups of O-H stretching of weak and intramolecular bonded. The strong band was appearing at 1725 cm$^{-1}$ and was generally ascribed to aldehyde groups of C=O stretching vibration [30]. The strong peak at 1530 cm$^{-1}$ was appearing and represent the nitro compound of N-O stretching vibration. The medium band at 1325 cm$^{-1}$ was observed and related to alcohol groups of O-H bending alcohol. The strong peak at 1110 cm$^{-1}$ was ascribed to the aliphatic ether of C-O stretching vibration. The two strong bands at 885 and 695 cm$^{-1}$ appearing may be related to alkene groups of C=C bending vibration [29].

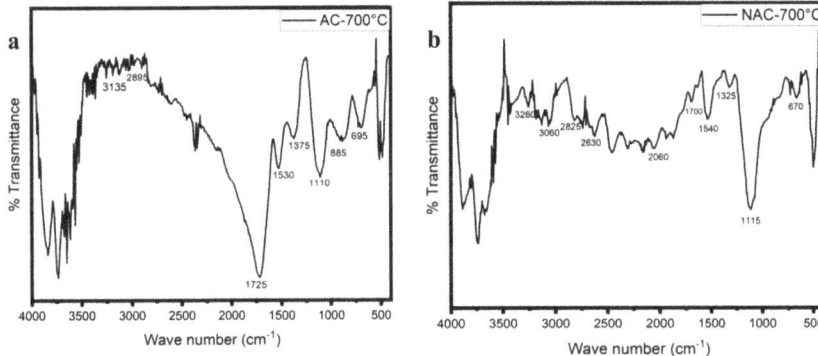

**Figure 4.** FTIR spectra of (**a**) AC-700°C and (**b**) NAC-700°C.

Figure 4b show the FTIR spectrum NAC-700 °C of the peak at 3260 cm$^{-1}$ was observed as a weak band and represents an alcohol group of O-H stretching vibration. The two small peaks around 3060 and 2825 cm$^{-1}$ were assigned to the aliphatic group CH, -CH$_2$, and –CH$_3$ [29,31]. The band at 2630 cm-1 was observed and assigned to aldehyde groups of C-H stretching. The weak band appearing at 1700 cm$^{-1}$ was allotted to carbonyl groups of C=O [29,32]. The strong peaks at 1540 cm$^{-1}$ and attributed to the nitro compound group of N-O stretching vibrations. The weak band at 1325 cm$^{-1}$ was generally credited to the alcohol group of O-H bending vibration. The strong peak at 1110 cm$^{-1}$ was ascribed to the aliphatic ether of C-O stretching vibration. The band at 670 cm$^{-1}$ may be ascribed to alkene groups of C=C bending vibration [29].

### 3.5. Cyclic Voltammetry (CV)

Evaluation of the capacitive performance of activated carbon prepared at various carbonization temperatures as electrode material for supercapacitor device was performed with cyclic voltammetry (CV) to study the result of scan rate on specific capacitance, using different scan rates (5–100 mV s$^{-1}$), and voltage windows of −1 to 0 V. As observed from Figure 5a shows EDLC behavior of AC and NAC as similar as ref [33], it is clearly seen the AC and NAC at 700 °C with 5 mV s$^{-1}$ shows a high enclosed area. In Figure 5b,c, The AC and NAC with the slight disorder and increasing the scan rate can change the area enclosed within the curve but decrease the specific capacitance value [23,25]. The best specific capacitance (196 F g$^{-1}$) resulted for AC-700 °C at the scan rate of 5 mV s$^{-1}$ in comparison to all other carbonization temperatures and NAC-700 °C has shown better specific capacitance (252 F g$^{-1}$) comparing to AC-700 °C from CV curves. Figure 5d shows that increasing the carbonization temperature guide to increasing the specific capacitance of the sample to an optimum temperature and then decreasing because this may decrease in the specific surface area directly impact specific capacitance due to high decomposition of material after optimum temperature [34]. These results show that AC-700 °C and NAC-700 °C exhibited the best performance and it is noteworthy, from Table 1 as the L$_c$ decreases there is an increase in SSA, so it leads to increase in specific capacitance and porosity [35,36].

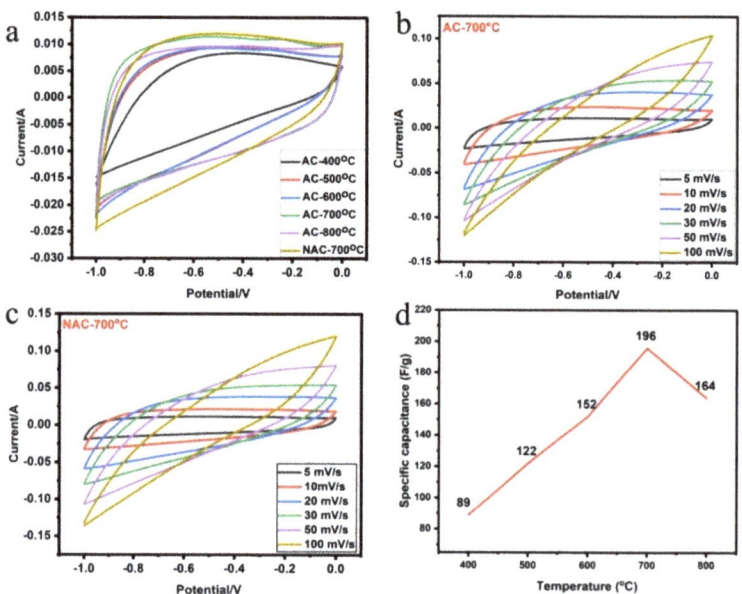

**Figure 5.** CV graph AC an NAC at various temperature (**a**) CV of AC and NAC at scan rate of 5 mV s$^{-1}$, (**b**) CV curve of AC-700 °C, (**c**) CV curve of NAC-700 °C, (**d**) Effect of temperature on specific capacitance from CV.

### 3.6. Galvano-Static Charge and Discharge (GCD)

The GCD curves have been performed for each sample (AC-400 °C to 800 °C and NAC-700 °C). Figure 6c,d shows the GCD curves for AC-700 °C and NAC-700 °C of supercapacitor devices with potential windows from −1.0 to 0 V at different current densities. The AC-700 °C and NAC-700 °C show the best electrochemical measurements compared to all other carbonization temperatures. Figure 6a shows GCD of AC and NAC at a current density of 0.25 A g$^{-1}$, It is note worthy that NAC shows a good dicscharge time. Figure 6d shows the specific capacitance of AC's is increased as with the carbonization temperature reaches up to 700 °C then start decreasing further due to the temperature rise leading to dissociation of the pore already existing [37]. Form Figure 6b,c AC-700 °C exhibits the specific capacitance of 183 F/g among all other carbonization temperature from the GCD curve and at this optimum carbonization temperature as prepared N-doped activated carbon (NAC-700 °C) shows a better specific capacitance of 258 F/g. After N-doping the improvement of specific capacitance is approximately 32%. The charge and discharge time decreases as the current densities increases because at low current density the charge transfer between the electron and electrolyte on the electrode surfaces is more [37]. The specific capacitance of all samples decreases as the current density increases. But at the same time, increase in current density didn't result in a change in the profile of the GCD graphs, revealing that both electrode material has high-rate capability. Thus, the two samples resulted in symmetric triangular shape. This quasi-symmetric triangular curve indicates the ideal properties of EDLC, high efficiency, and good reversibility in the charge-discharge process [23,38]. For comparative studies, the NAC performance as electrode materials was analyzed with other N-doped activated carbon as listed in Table 2.

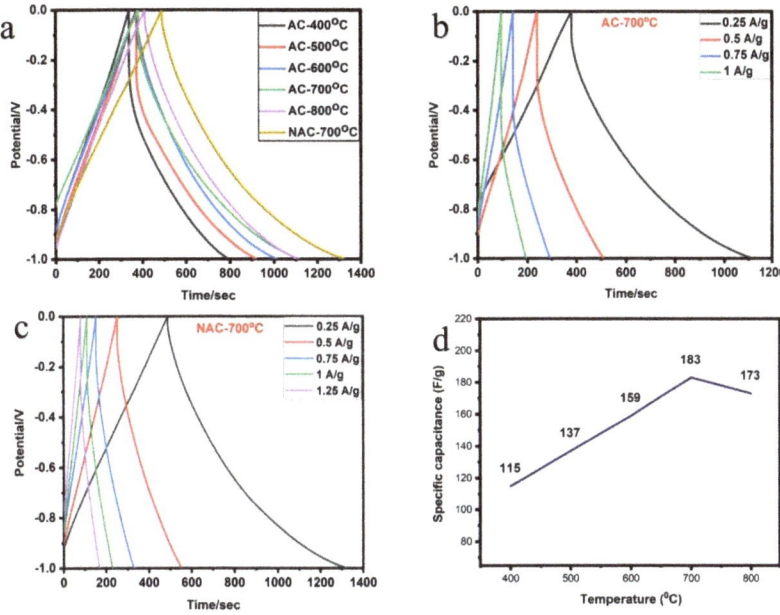

**Figure 6.** GCD graph of AC an NAC at various temperature (**a**) GCD of AC and NAC at a current density of 0.25 A g$^{-1}$, (**b**) GCD curve of AC-700 °C, (**c**) GCD curve of NAC-700 °C, (**d**) effect of temperature on specific capacitance from GCD.

**Table 2.** Comparison of electrochemical performance NAC derived from different precursors.

| Raw Materials | Precursors | Electrolyte | Specific Capacitance (F g$^{-1}$) | Energy Density (Wh kg$^{-1}$) | Ref. |
|---|---|---|---|---|---|
| Orange peel | KOH + Melamine | 6 M KOH | 168 | 23.3 | [39] |
| corncob | KOH + NH$_3$ | Organic | 185 | - | [40] |
| Pea skin | KOH + Melamine | 1 M LiTFSI in 1 L EMITFSI | 141 | 19.6 | [41] |
| Peony pollen | KOH + NH$_4$[BF$_4$] | 6 M KOH | 209 | - | [42] |
| Macadamia nutshell | KOH + Melamine | 1 M KOH | 229 | - | [43] |
| Glucosamine | - | PVA/ KOH | 244 | 7.2 | [44] |
| Tea seed shell | KOH + Melamine | 1 M KOH | 141 | - | [45] |
| Pueraria | Melamine + K$_2$CO$_3$ | 6 M KOH | 250 | 8.46 | [46] |
| Eucalyptus Leaves | ZnCl$_2$ + Urea | 2 KOH | 258 | 28.76 | Present work |

## 4. Conclusions

In summary, activated carbon (AC) and N-Doped activated carbon (NAC) were synthesized from eucalyptus leaves using chemical activation techniques at various temperature with ZnCl$_2$ as activator, and the source of nitrogen was loaded using urea. The synthesized AC at 700 °C showed a better performance with a surface area of 1027 m$^2$ g$^{-1}$ and the enhanced performance was observed in NAC at 700 °C with a surface area of have a surface area 1042 m$^2$ g$^{-1}$. The impact of activator and nitrogen doped showed different morphology such as porous, graphitic sheet, honeycomb structure. The specific capacitance of AC and NAC were obtained 194 F g$^{-1}$ and 258 F g$^{-1}$, respectively, at a scan rate of 5 mV s$^{-1}$. The energy densities of AC and NAC were obtained 22 Wh Kg$^{-1}$ and 28.76 Wh

Kg$^{-1}$, respectively, at a current density of 0.25 A g$^{-1}$, for the carbon electrode. This carbon electrode shows good performance with alkaline electrolyte. The electrochemical performance suggest that AC and NAC were potentially excellent electrode for supercapacitor cell application. Finally, the findings of this study highlight the need of defining abundant residue as a necessary step in determining prospective usage for a subsequent valorization.

**Supplementary Materials:** The following supporting information can be downloaded at: https://www.mdpi.com/article/10.3390/c9010024/s1, Figure S1: EDS spectra; Table S1: EDS summary.

**Author Contributions:** D.B.: Experimental, results, investigation, writing original draft, review, and revisions. P.B.: Experimental, results analysis, revision, and editing. V.B.K.: conceptualization review and editing. P.S.K.: Supervision; conceptualization, review, suggestions, and editing. All authors have read and agreed to the published version of the manuscript.

**Funding:** This research received no external funding.

**Data Availability Statement:** Not applicable.

**Conflicts of Interest:** The authors declare no conflict of interest.

## References

1. Amirante, R.; Cassone, E.; Distaso, E.; Tamburrano, P. Overview on Recent Developments in Energy Storage: Mechanical, Electrochemical and Hydrogen Technologies. *Energy Convers. Manag.* **2017**, *132*, 372–387. [CrossRef]
2. Aliyu, A.K.; Modu, B.; Tan, C.W. A Review of Renewable Energy Development in Africa: A Focus in South Africa, Egypt and Nigeria. *Renew. Sustain. Energy Rev.* **2018**, *81*, 2502–2518. [CrossRef]
3. Tian, J.; Zhang, T.; Talifu, D.; Abulizi, A.; Ji, Y. Porous Carbon Materials Derived from Waste Cotton Stalk with Ultra-High Surface Area for High Performance Supercapacitors. *Mater. Res. Bull.* **2021**, *143*, 111457. [CrossRef]
4. Dubey, P.; Shrivastav, V.; Maheshwari, P.H.; Sundriyal, S. Recent Advances in Biomass Derived Activated Carbon Electrodes for Hybrid Electrochemical Capacitor Applications: Challenges and Opportunities. *Carbon* **2020**, *170*, 1–29. [CrossRef]
5. Chen, T.; Dai, L. Carbon Nanomaterials for High-Performance Supercapacitors. *Mater. Today* **2013**, *16*, 272–280. [CrossRef]
6. Atika; Dutta, R.K. Oxygen-Rich Porous Activated Carbon from *Eucalyptus* Wood as an Efficient Supercapacitor Electrode. *Energy Technol.* **2021**, *9*, 2100463. [CrossRef]
7. Stoller, M.D.; Ruoff, R.S. Best Practice Methods for Determining an Electrode Material's Performance for Ultracapacitors. *Energy Environ. Sci.* **2010**, *3*, 1294–1301. [CrossRef]
8. Tan, Y.; Li, Y.; Wang, W.; Ran, F. High Performance Electrode of Few-Layer-Carbon@bulk-Carbon Synthesized via Controlling Diffusion Depth from Liquid Phase to Solid Phase for Supercapacitors. *J. Energy Storage* **2020**, *32*, 101672. [CrossRef]
9. Zhang, L.; Zhao, X.S. Carbon-Based Materials as Supercapacitor Electrodes. *Chem. Soc. Rev.* **2009**, *38*, 2520–2531. [CrossRef]
10. Najib, S.; Erdem, E. Current Progress Achieved in Novel Materials for Supercapacitor Electrodes: Mini Review. *Nanoscale Adv.* **2019**, *1*, 2817–2827. [CrossRef]
11. Li, Z.; Guo, D.; Liu, Y.; Wang, H.; Wang, L. Recent Advances and Challenges in Biomass-Derived Porous Carbon Nanomaterials for Supercapacitors. *Chem. Eng. J.* **2020**, *397*, 125418. [CrossRef]
12. Sellin, N.; Krohl, D.R.; Marangoni, C.; Souza, O. Oxidative Fast Pyrolysis of Banana Leaves in Fluidized Bed Reactor. *Renew. Energy* **2016**, *96*, 56–64. [CrossRef]
13. Pujol, D.; Liu, C.; Gominho, J.; Olivella, M.À.; Fiol, N.; Villaescusa, I.; Pereira, H. The Chemical Composition of Exhausted Coffee Waste. *Ind. Crops Prod.* **2013**, *50*, 423–429. [CrossRef]
14. Al-Maliki, S.; Al-Masoudi, M. Interactions between Mycorrhizal Fungi, Tea Wastes, and Algal Biomass Affecting the Microbial Community, Soil Structure, and Alleviating of Salinity Stress in Corn Yield (*Zea mays* L.). *Plants* **2018**, *7*, 63. [CrossRef]
15. Liu, Y.; Wang, Y.; Zhang, G.; Liu, W.; Wang, D.; Dong, Y. Preparation of Activated Carbon from Willow Leaves and Evaluation in Electric Double-Layer Capacitors. *Mater. Lett.* **2016**, *176*, 60–63. [CrossRef]
16. Guo, N.; Li, M.; Sun, X.; Wang, F.; Yang, R. Tremella Derived Ultrahigh Specific Surface Area Activated Carbon for High Performance Supercapacitor. *Mater. Chem. Phys.* **2017**, *201*, 399–407. [CrossRef]
17. El-Sayed, S.A.; Mostafa, M.E. Thermal Pyrolysis and Kinetic Parameter Determination of Mango Leaves Using Common and New Proposed Parallel Kinetic Models. *RSC Adv.* **2020**, *10*, 18160–18179. [CrossRef]
18. Grima-Olmedo, C.; Ramírez-Gómez, A.; Gómez-Limón, D.; Clemente-Jul, C. Activated Carbon from Flash Pyrolysis of *Eucalyptus* Residue. *Heliyon* **2016**, *2*, e00155. [CrossRef]
19. Gan, Y.X. Activated Carbon from Biomass Sustainable Sources. *C* **2021**, *7*, 39. [CrossRef]
20. Roy, C.K.; Shah, S.S.; Reaz, A.H.; Sultana, S.; Chowdhury, A.N.; Firoz, S.H.; Zahir, M.H.; Ahmed Qasem, M.A.; Aziz, M.A. Preparation of Hierarchical Porous Activated Carbon from Banana Leaves for High-Performance Supercapacitor: Effect of Type of Electrolytes on Performance. *Chem. Asian J.* **2021**, *16*, 296–308. [CrossRef]

21. Bergna, D.; Varila, T.; Romar, H.; Lassi, U. Comparison of the Properties of Activated Carbons Produced in One-Stage and Two-Stage Processes. *C* **2018**, *4*, 41. [CrossRef]
22. Zhu, X.; Yu, S.; Xu, K.; Zhang, Y.; Zhang, L.; Lou, G.; Wu, Y.; Zhu, E.; Chen, H.; Shen, Z.; et al. Sustainable Activated Carbons from Dead Ginkgo Leaves for Supercapacitor Electrode Active Materials. *Chem. Eng. Sci.* **2018**, *181*, 36–45. [CrossRef]
23. Sylla, N.F.; Ndiaye, N.M.; Ngom, B.D.; Mutuma, B.K.; Momodu, D.; Chaker, M.; Manyala, N. Ex-Situ Nitrogen-Doped Porous Carbons as Electrode Materials for High Performance Supercapacitor. *J. Colloid Interface Sci.* **2020**, *569*, 332–345. [CrossRef] [PubMed]
24. Xiao, P.W.; Meng, Q.; Zhao, L.; Li, J.J.; Wei, Z.; Han, B.H. Biomass-Derived Flexible Porous Carbon Materials and Their Applications in Supercapacitor and Gas Adsorption. *Mater. Des.* **2017**, *129*, 164–172. [CrossRef]
25. Farma, R.; Kusumasari, M.; Apriyani, I.; Awitdrus, A. The Production of Carbon Electrodes from Lignocellulosic Biomass of *Areca Midrib* through a Chemical Activation Process for Supercapacitor Cells Application. *Energy Sources A Recover. Util. Environ. Eff.* **2021**, 1–11. [CrossRef]
26. Lee, S.M.; Lee, S.H.; Roh, J.S. Analysis of Activation Process of Carbon Black Based on Structural Parameters Obtained by XRD Analysis. *Crystals* **2021**, *11*, 153. [CrossRef]
27. Pang, X.; Cao, M.; Qin, J.; Li, X.; Yang, X. Synthesis of Bamboo-Derived Porous Carbon: Exploring Structure Change, Pore Formation and Supercapacitor Application. *J. Porous Mater.* **2022**, *29*, 559–569. [CrossRef]
28. Ratnaji, T.; Kennedy, L.J. Hierarchical Porous Carbon Derived from Tea Waste for Energy Storage Applications: Waste to Worth. *Diam. Relat. Mater.* **2020**, *110*, 108100. [CrossRef]
29. Benadjemia, M.; Millière, L.; Reinert, L.; Benderdouche, N.; Duclaux, L. Preparation, Characterization and Methylene Blue Adsorption of Phosphoric Acid Activated Carbons from Globe Artichoke Leaves. *Fuel Process. Technol.* **2011**, *92*, 1203–1212. [CrossRef]
30. Xu, J.; Chen, L.; Qu, H.; Jiao, Y.; Xie, J.; Xing, G. Preparation and Characterization of Activated Carbon from Reedy Grass Leaves by Chemical Activation with $H_3PO_4$. *Appl. Surf. Sci.* **2014**, *320*, 674–680. [CrossRef]
31. Saka, C. BET, TG-DTG, FT-IR, SEM, Iodine Number Analysis and Preparation of Activated Carbon from Acorn Shell by Chemical Activation with $ZnCl_2$. *J. Anal. Appl. Pyrolysis* **2012**, *95*, 21–24. [CrossRef]
32. Joshi, S.; Pokharel, B.P. Preparation and Characterization of Activated Carbon from Lapsi (*Choerospondias axillaris*) Seed Stone by Chemical Activation with Potassium Hydroxide. *J. Inst. Eng.* **2014**, *9*, 79–88. [CrossRef]
33. Tsyganova, S.; Mazurova, E.; Bondarenko, G.; Fetisova, O.; Skvortsova, G. Structural and Current-Voltage Characteristics of Carbon Materials Obtained during Carbonization of Fir and Aspen Barks. *Biomass Bioenergy* **2020**, *142*, 105759. [CrossRef]
34. Ochai-Ejeh, F.O.; Bello, A.; Dangbegnon, J.; Khaleed, A.A.; Madito, M.J.; Bazegar, F.; Manyala, N. High Electrochemical Performance of Hierarchical Porous Activated Carbon Derived from Lightweight Cork (*Quercus suber*). *J. Mater. Sci.* **2017**, *52*, 10600–10613. [CrossRef]
35. Sarkar, J.; Bhattacharyya, S. Application of Graphene and Graphene-Based Materials in Clean Energy-Related Devices Minghui. *Arch. Thermodyn.* **2012**, *33*, 23–40. [CrossRef]
36. Puttapati, S.K.; Gedela, V.; Srikanth, V.V.S.S.; Reddy, M.V.; Adams, S.; Chowdari, B.V.R. Unique Reduced Graphene Oxide as Efficient Anode Material in Li Ion Battery. *Bull. Mater. Sci.* **2018**, *41*, 53. [CrossRef]
37. Farma, R.; Apriyani, I.; Awitdrus, A.; Taer, E.; Apriwandi, A. Hemicellulosa-Derived *Arenga pinnata* Bunches as Free-Standing Carbon Nanofiber Membranes for Electrode Material Supercapacitors. *Sci. Rep.* **2022**, *12*, 2572. [CrossRef]
38. Gou, H.; He, J.; Zhao, G.; Zhang, L.; Yang, C.; Rao, H. Porous Nitrogen-Doped Carbon Networks Derived from Orange Peel for High-Performance Supercapacitors. *Ionics* **2019**, *25*, 4371–4380. [CrossRef]
39. Ahmed, S.; Rafat, M.; Ahmed, A. Nitrogen Doped Activated Carbon Derived from Orange Peel for Supercapacitor Application. *Adv. Nat. Sci. Nanosci. Nanotechnol.* **2018**, *9*, 035008. [CrossRef]
40. Li, B.; Dai, F.; Xiao, Q.; Yang, L.; Shen, J.; Zhang, C.; Cai, M. Nitrogen-Doped Activated Carbon for a High Energy Hybrid Supercapacitor. *Energy Environ. Sci.* **2016**, *9*, 102–106. [CrossRef]
41. Ahmed, S.; Ahmed, A.; Rafat, M. Nitrogen Doped Activated Carbon from Pea Skin for High Performance Supercapacitor. *Mater. Res. Express* **2018**, *5*, 045508. [CrossRef]
42. Liu, Y.; An, Z.; Wu, M.; Yuan, A.; Zhao, H.; Zhang, J.; Xu, J. Peony Pollen Derived Nitrogen-Doped Activated Carbon for Supercapacitor Application. *Chin. Chem. Lett.* **2020**, *31*, 1644–1647. [CrossRef]
43. Li, Z.; Liang, Q.; Yang, C.; Zhang, L.; Li, B.; Li, D. Convenient Preparation of Nitrogen-Doped Activated Carbon from Macadamia Nutshell and Its Application in Supercapacitor. *J. Mater. Sci. Mater. Electron.* **2017**, *28*, 13880–13887. [CrossRef]
44. Xu, Z.; Chen, J.; Zhang, X.; Song, Q.; Wu, J.; Ding, L.; Zhang, C.; Zhu, H.; Cui, H. Template-Free Preparation of Nitrogen-Doped Activated Carbon with Porous Architecture for High-Performance Supercapacitors. *Microporous Mesoporous Mater.* **2019**, *276*, 280–291. [CrossRef]

45. Quan, C.; Jia, X.; Gao, N. Nitrogen-Doping Activated Biomass Carbon from Tea Seed Shell for $CO_2$ Capture and Supercapacitor. *Int. J. Energy Res.* **2020**, *44*, 1218–1232. [CrossRef]
46. Han, X.; Jiang, H.; Zhou, Y.; Hong, W.; Zhou, Y.; Gao, P.; Ding, R.; Liu, E. A High Performance Nitrogen-Doped Porous Activated Carbon for Supercapacitor Derived from *Pueraria*. *J. Alloys Compd.* **2018**, *744*, 544–551. [CrossRef]

**Disclaimer/Publisher's Note:** The statements, opinions and data contained in all publications are solely those of the individual author(s) and contributor(s) and not of MDPI and/or the editor(s). MDPI and/or the editor(s) disclaim responsibility for any injury to people or property resulting from any ideas, methods, instructions or products referred to in the content.

*Article*

# Active Carbon-Based Electrode Materials from Petroleum Waste for Supercapacitors

Abdualilah Albaiz, Muhammad Alsaidan, Abdullah Alzahrani, Hassan Almoalim, Ali Rinaldi *,†
and Almaz S. Jalilov *

Department of Chemistry and Interdisciplinary Research Center for Advanced Materials, King Fahd University of Petroleum and Minerals, Dhahran 31261, Saudi Arabia
* Correspondence: ali.rinaldi@tum.de (A.R.); jalilov@kfupm.edu.sa (A.S.J.)
† Current Address: Faculty of Chemistry, School of Natural Sciences, Technical University of Munich, Lichtenbergstrasse 4, 85748 Garching bei München, Germany.

**Abstract:** A supercapacitor is an energy-storage device able to store and release energy at fast rates with an extended cycle life; thus, it is used in various electrical appliances. Carbon materials prepared above 800 °C of activation temperatures are generally employed as an electrode material for supercapacitors. Herein, we report carbon materials prepared from a low-cost petroleum waste carbon precursor that was activated using KOH, MgO, and Ca(OH)$_2$ only at 400 °C. Electrode materials using low-temperature activated carbons were prepared with commercial ink as a binder. The cyclic voltammetry and galvanostatic charge–discharge were employed for the electrochemical performance of the electrodes, and studied in a 3-electrode system in 1 M solutions of potassium nitrate (KNO$_3$) as electrolyte; in addition, the supercapacitive performance was identified in a potential window range of 0.0–1.0 V. The best-performance activated carbon derived from vacuum residue with a specific surface area of 1250.6 m$^2$/g exhibited a specific capacitance of 91.91 F/g.

**Keywords:** activated carbon; vacuum residue; supercapacitor; carbon nanotechnology

---

**Citation:** Albaiz, A.; Alsaidan, M.; Alzahrani, A.; Almoalim, H.; Rinaldi, A.; Jalilov, A.S. Active Carbon-Based Electrode Materials from Petroleum Waste for Supercapacitors. *C* **2023**, *9*, 4. https://doi.org/10.3390/c9010004

Academic Editors: Indra Pulidindi, Pankaj Sharma and Aharon Gedanken

Received: 10 November 2022
Revised: 17 December 2022
Accepted: 22 December 2022
Published: 28 December 2022

**Copyright:** © 2022 by the authors. Licensee MDPI, Basel, Switzerland. This article is an open access article distributed under the terms and conditions of the Creative Commons Attribution (CC BY) license (https:// creativecommons.org/licenses/by/ 4.0/).

## 1. Introduction

Supercapacitors, i.e., electrochemical capacitors, have a high-power density, fast charging rate, cyclability, and low cost of operation [1,2]. Thus, they have found many uses in electronics, energy management, mobile electrical systems, as well as industrial power arrays [3,4]. In an electrochemical double-layer supercapacitor, the energy is stored in the electrostatic charge separation on the surface between the electrode and electrolyte. High porosity and good mechanical properties allow porous carbon materials to be widely used as electrode materials [5].

In a supercapacitor, energy is stored either by an electrochemical double layer in the electrode–electrolyte interphase (nonfaradaic) and/or by reduction–oxidation reactions (faradaic) on electrode surfaces [6–12]. For electrochemical double-layer capacitors (EDLC), ions in the electrolyte will accumulate at the surface of the solid electrode materials. The capacitance (CDL) of the electrode depends on the layer thickness where the ions and solvent molecules reside, d; and the surface area of the electrode, A (see equation below). $\varepsilon$ and $\varepsilon r$ are the permittivity of the vacuum and the solvent, respectively.

$$CDL = \varepsilon \times \varepsilon r\, A/d$$

For high-energy and power-density EDLC, it is imperative for the electrode to possess a high surface area for cations and anions to accumulate. Porous carbons, such as graphene-based nanocarbons, carbon nanotubes, and activated carbons, are used as electrodes in EDLC supercapacitor devices due to their high surface area, high electrical conductivity, and electrochemical stability. The low-cost and established electrode fabrication technologies in the industry create a sustainable approach for using carbons as electrodes in supercapacitor devices. Activated carbon with a BET surface area of 3150 m$^2$/g was reported to show a capacitance of 312 F/g. This capacitance translates into a specific capacitance value of 9.9 µF/cm$^2$ [13].

Porous carbons from various synthesis routes have been explored as high-surface area electrodes for supercapacitors [14,15]. Asphalt, a petroleum waste, has been used as a precursor to high-surface area carbons that were used as electrodes in batteries and supercapacitors [16–18]. Asphalt is a low-cost material obtained from the heaviest fraction of crude oil. Asphalt contains some amount of volatile organic species that are removed during a carbonization process at about >400 °C. Asphalt-derived carbons with surface areas of >4000 m$^2$/g were achieved after a sequential process of removal of volatile organic followed by high-temperature activation with KOH [19,20].

This energy-intensive activation step to create a high surface area can be compensated by using alternative and inexpensive activating agents from industrial waste. Coal-fired power plants, on the other hand, are the large-point sources of carbon dioxide ($CO_2$) [21] and generate solid waste such as fly ash and bottom ash, which mainly contain lime (CaO) and magnesium oxide (MgO) [22,23]. However, using such alternative and inexpensive mineral feedstock from industrial waste as activating agents is challenging due to the slow chemical kinetics and high activation energy [10]. Furthermore, most of the methods of activation of carbons require high temperatures, usually above 800 °C. The combination of using asphalts as petroleum waste as a carbon source and solid industrial waste as activating agents can be an attractive cost-effective route towards synthesizing high-surface-area carbons for supercapacitor application.

In this work, we aim to use vacuum residue (VR) as the carbon precursor for generating low-density porous carbon materials with KOH, MgO, and Ca(OH)$_2$ activation under mild conditions of activation at 400 °C. The resultant activated carbon materials will be tested as electrode materials for supercapacitors.

## 2. Materials and Methods

Materials. Vacuum residue (VR) was acquired from the local Ras Tanura refinery in the Eastern Province of Saudi Arabia. The average composition of VR is >85 wt% carbon, ~10 wt% hydrogen, ~4 wt% sulfur, and <1 wt% nitrogen; Vulcan black, calcium hydroxide (Ca(OH)$_2$), magnesium oxide (MgO), and potassium hydroxide (KOH) were purchased from Millipore-Sigma and used without further purification; ultrapure water processed from the Milli-Q (Milford, MA, USA) system.

Synthesis. The preparation of porous carbon materials from VR was conducted in one step using activating agents, such as calcium hydroxide (Ca(OH)$_2$), magnesium oxide (MgO), and potassium hydroxide (KOH). Briefly, different ratios of VR and the activation agents are mixed and placed in a horizontal tubular furnace and heat treated under an inert atmosphere at 400 °C for 4 h. The activation procedures are employed with the ratio = 1:4 (between VR: activating agent), denoted as VR-KOH, VR-Ca(OH)$_2$, VR-MgO. After the activation, the samples are washed with deionized water until a neutral pH and oven-dried at 110 °C overnight.

Characterization. X-ray photoelectron spectroscopic (XPS) analyses were obtained on a Thermo Scientific Escalab 250Xi spectrometer with Al Kα (1486.6 eV) as the x-ray source and an operating resolution of 0.5 eV. X-rays with a 650 µm beam and pass energy of 100 eV are used for the survey scan, and 30 eV is used for the high-resolution scans. High-resolution spectra for binding energies spectra were centered at 284.8 eV, corresponding to the C 1s of the graphitic carbon (C–C/C=C). SDT Q600 (TA instruments) was used for the thermal gravimetric analysis (TGA). Typically, the measurement was performed by heating ~10 mg weight of samples up to 900 °C in aluminum pans at a constant heating rate of 10 °C min$^{-1}$ under N$_2$ flow (99.999% purity). A Quattro ESEM 400 high-resolution field emission scanning electron microscope (SEM) at 20 keV is used for the SEM images and energy dispersive x-ray analysis (EDX). Brunauer–Emmett–Teller (BET) analysis was performed to investigate the texture of the carbon materials using the Quantachrome Autosorb-

3b. Prior to the $N_2$ physisorption measurements, the samples were activated at 300 °C for 24 h under a vacuum.

Electrode Preparation. Graphite foil was cut in circular form with a diameter of 1 cm and used as the working electrode. Typically, 1 gr of ink is mixed with the ~30 mg of active sample and the suspension is sonicated for 10 min. In all, ~20 mg of well-dispersed suspension is drop-casted on the graphite disc and dried at 85 °C for 48 h.

## 3. Results and Discussion

To examine the efficiency of the activating agents and the composition of the final materials, TGA was performed under air for all the materials after the activation step. It is expected that under an air atmosphere, the carbon content will be combusted leaving only the metal oxides. VR-KOH exhibited a major weight-loss profile at 350–470 °C, which corresponds to 93% weight of its original weight. This step is due to the combustion of carbon as shown in Figure 1b. On the other hand, both VR-Ca(OH)$_2$ and VR-MgO exhibit distinct thermal behavior as shown in Figure 1c,d, respectively. The major weight loss step for VR-Ca(OH)$_2$ occurred at 600–710 °C, corresponding to the combustion of carbon components in 32% of total weight. The remaining 68% is due to the CaO content. The thermal weight loss profile for VR-MgO reveals the major weight loss step at 350–500 °C, due to the combustion of carbon. This constitutes 17% of the total weight, pointing to 83% of the MgO composition in VR-MgO (Figure 1d).

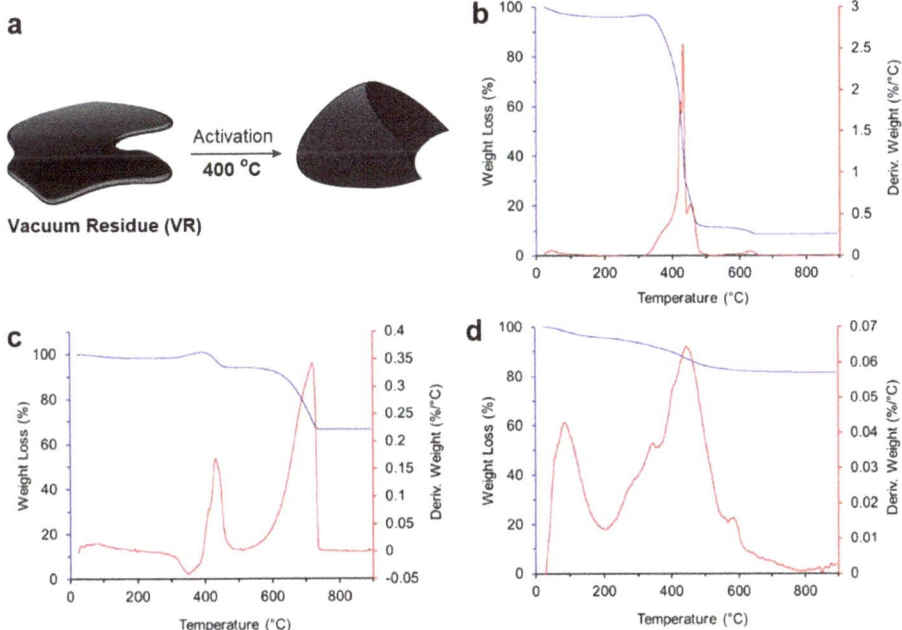

**Figure 1.** (**a**) Schematic representation of petroleum waste activation for porous activated carbon materials. Thermogravimetric curves of (**b**) VR-KOH, (**c**) VR-Ca(OH)$_2$, and (**d**) VR-MgO (the heating rate at 10 °C min$^{-1}$, under air).

The morphology of the activated carbons was investigated using SEM using either SE or BSE signals. In general, the morphology of the activated carbons herein is different than carbon black, where carbon nanoparticles form an aggregate. Carbon samples activated with either KOH, Ca(OH)$_2$, or MgO exhibit irregular shapes of micron-sized particles (Figure 2).

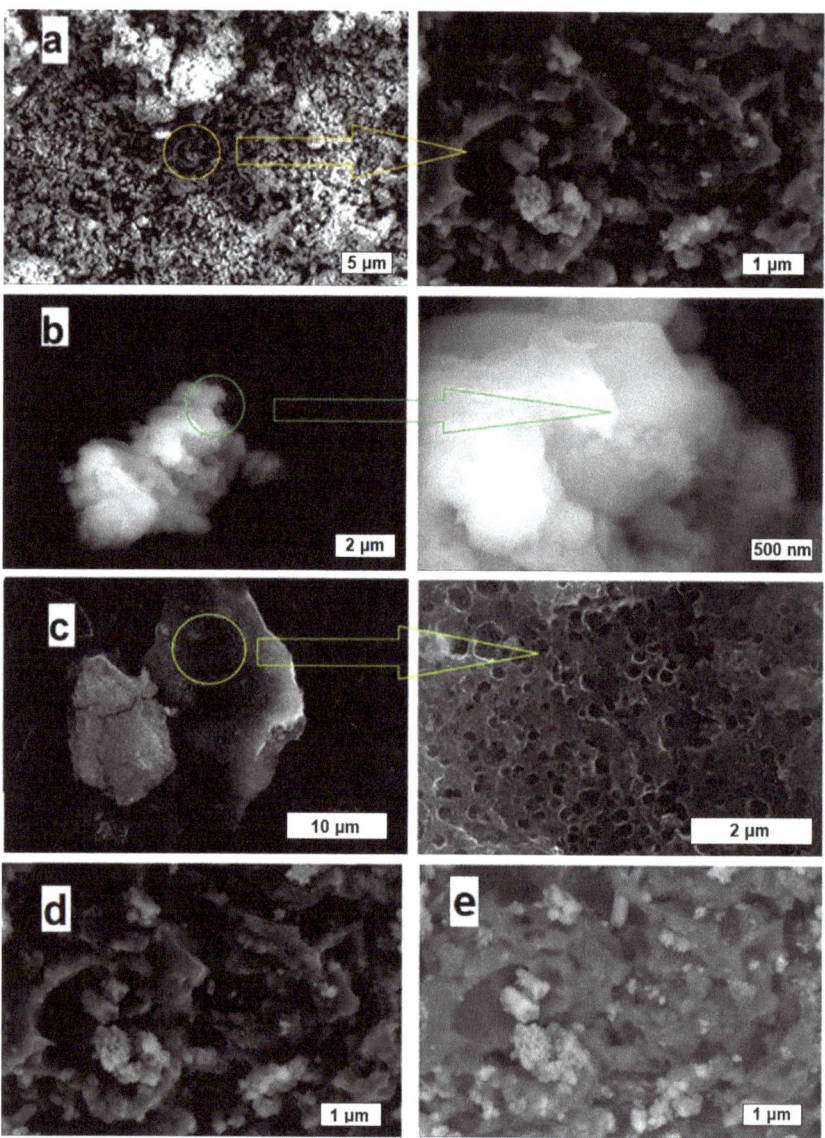

**Figure 2.** SEM images of (**a**) VR-KOH, (**b**) VR-Ca(OH)$_2$, (**c**) VR-MgO; and (**d**) SE and (**e**) BSE imaging mode of VR-KOH.

SEM analysis for VR-KOH (Figure 2a) shows a broad range of particle size of about 0.5 to 5 um. Irregular-shaped cavities were observed on the particles with sizes of about 50 to 500 nm. Inorganic materials were also observed in the sample dispersed on the carbon particle surfaces and cavities as individual particles or agglomerates. These inorganic materials on the carbon sample can be seen from their brighter contrast in the BSE and SE imaging modes (additional SEM images). It is safe to say that the inorganic particles are mainly composed of KOH, the activating agent added to the asphalt precursor. This agrees with the TGA measurement, which revealed a high carbon content in VR-KOH. Due to the low carbon content of the VR-Ca(OH)$_2$ sample and the non-conducting nature of Ca(OH)$_2$ particles, SEM imaging analyses were performed with a low-vacuum mode under water vapor. Nevertheless, VR-Ca(OH)$_2$ shows agglomerated nanoparticles of mainly CaO, with sizes in the range of 50–500 nm (Figure 2b). The morphology of the VR-MgO sample shows a greater amount of macropores compared to the other two activated samples as shown in Figure 2c. The macropores present in VR-MgO are in the range of 100 nm to 5 um. The porosity produced on the activated sample depends on the activating agent. This indicates the different activating or oxidizing power of the activating agents. In addition, other factors such as the initial dispersion of the activating agents on asphalt may also lead to the varying progression of pore formation. The EDX measurements at 20 kV addressed the EDX-average value of the samples' sulfur content. The measurements on all of the activated samples show a low content of sulfur. The VR-MgO sample showed the highest sulfur content around 1.0 wt%, while both the VR-KOH and VR-Ca(OH)$_2$ samples show less than 0.1 wt%. EDX measurements were performed with a minimum of five spots for every sample.

A more macroscopic measurement of the activated samples' texture was obtained from N$_2$ adsorption isotherms at 77 K (Figure 3). The porosity is generated as a result of asphalt (vacuum residue) oxidation by the chemical activating agents under an inert atmosphere. A carbon black sample (Vulcan black) adsorption isotherm was used as a benchmark to compare adsorption isotherms of the VR-derived carbons. The carbon black sample has no microporosity and has only mesopores due to spaces between the carbon black nanoparticles. The BET surface areas were calculated from the adsorption isotherms as a measure of the total surface areas of the samples. The BET total surface area is the sum of the external surface area and the micropore surface area. The t-plot method was used to determine the external surface area (Table 1). In the t-plot method, the statistical thickness of the adsorbed N$_2$ onto the carbon surface is calculated as a function of the adsorbed volume. More adsorption at a higher P/Po will give higher thickness. The deviation will from linearity corresponds to micropore filling. The micropore area was then calculated by subtracting the BET total area from the external area. The total pore volume was calculated from the amount of N$_2$ adsorbed at a relative pressure of P/Po = 0.99. This pore volume corresponds to micro and mesopores filled until the N$_2$ gas reaches full condensation on the sample. The carbon black reference VR-Ca(OH)$_2$ and VR-MgO samples exhibit a similar adsorption isotherm feature with insignificant N$_2$ adsorption at P/Po < 0.01. This indicates a negligible amount of micropores. On the contrary, the VR-KOH sample shows appreciable adsorption at P/Po < 0.01, indicating abundant microporosity. Further analysis of the isotherms shows that the VR-KOH sample has a high BET total surface area and high micropore area. The external surface areas of the activated samples are of the same order of magnitude with some variation. However, the VR-MgO sample has about 40% more external area than the other two activated samples. This is supported by the more abundant large pores in the VR-MgO sample as seen in the SEM images (Figure 2c). KOH is better than either LiOH and NaOH in producing microporosity at 400 °C than either Ca(OH)$_2$ or MgO.

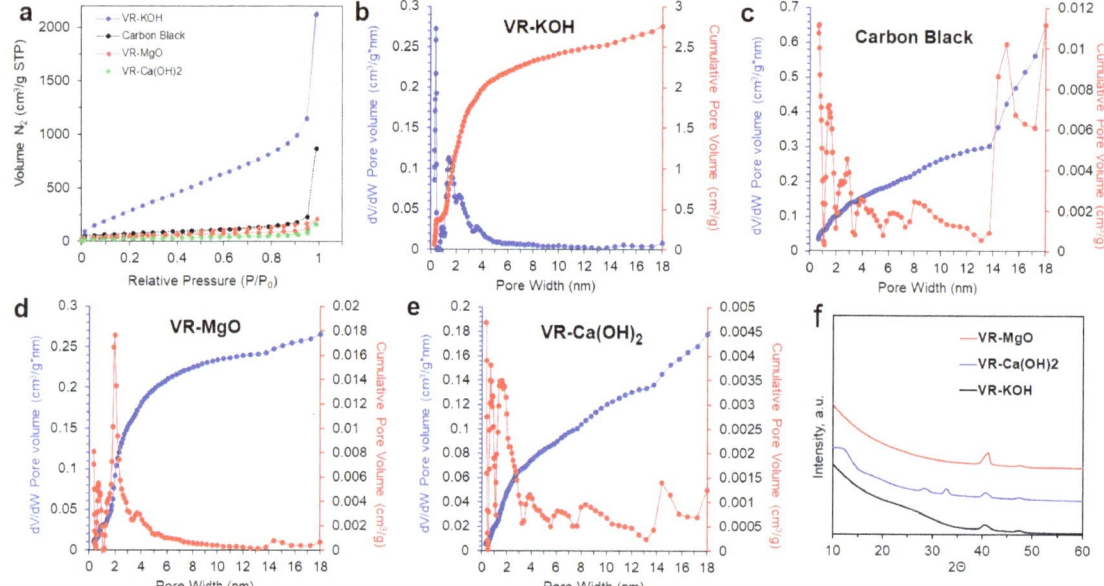

**Figure 3.** (**a**) Comparative $N_2$-adsorption isotherms classified as type I for VR-KOH, carbon black, VR-MgO, and VR-Ca(OH)$_2$; pore-size distribution curves for (**b**) VR-KOH, (**c**) carbon black, (**d**) VR-MgO, and (**e**) VR-Ca(OH)$_2$ determined using the NLDFT and (**f**) XRD pattern of VR-KOH, VR-MgO, and VR-Ca(OH)$_2$.

Textural parameters and specific surface areas for all the samples are summarized in Table 1. The $N_2$-adsorption isotherms are also shown in Figure 3 concerning a commercial carbon black. Micro- and mesoporous structures of the samples were also evident from the pore size distributions obtained by applying the NLDFT depicted in Figure 3b–e. VR-KOH and VR-MgO possess pore sizes in the range of 1–8 nm. On the other hand, a much wider pore size distribution range of up to 18 nm was observed for both VR-Ca(OH)$_2$ and carbon black. The XRD pattern shown in Figure 3f, reveals the relatively intense broad peak at the lower angle region of ~25 for VR-KOH, indicating a higher degree of graphitization in comparison to both VR-MgO and VR-Ca(OH)$_2$. All the samples possess the peak at ~40 assigned for (10) band reflections coming from the polyaromatic structure of asphaltenes, indicating a not complete carbonization of the VR at 400 °C activations. VR-Ca(OH)$_2$ also shows peaks at ~28 and ~33, demonstrating the presence of crystalline CaO.

**Table 1.** $N_2$-physisorption analysis data.

| Sample | $S_{BET}$ [1,5] [m$^2$ g$^{-1}$] | $V_{total}$ [2] [cm$^3$ g$^{-1}$] | Micropore Area [3] [m$^2$ g$^{-1}$] | External Surface Area [4] [m$^2$ g$^{-1}$] |
|---|---|---|---|---|
| VR-KOH | 1250.6 | 5.3 | 1146 | 103.9 |
| VR-Ca(OH)$_2$ | 99.8 | 0.259 | 5.33 | 94.47 |
| VR-MgO | 179.8 | 0.193 | 36.3 | 143.5 |
| Carbon black | 266.9 | 0.363 | 50.2 | 216 |

[1] BET area can be considered as the total surface area. [2] Taken at P/Po = 0.99. [3] Pore with diameter < 2 nm, when there is microporosity; BET area = micropore area + external surface area (t-plot). [4] Pore with diameter > 2 nm. [5] Repeated measurements of representative sample show about 14% standard deviation of BET values.

XPS analysis of the chemical surface composition of materials is summarized in Figure 4 and Table 2. Surface characterization of VR-KOH reveals 84.23% of carbon and 11.85% oxygen content with the amount of sulfur present at 2.09% and nitrogen at 1.83%. VR-Ca(OH)$_2$ possesses 42.75% of carbon, 39.09% of oxygen, and 16.85% calcium, with the amount of nitrogen at 1.32%. VR-MgO possesses 34.09% of carbon, 52.69% of oxygen, and 13.22% magnesium. High-resolution C 1s XPS spectra with fitted deconvolution of the peaks for VR-KOH show the majority of carbon is in aromatic and aliphatic carbon form, whereas both VR-Ca(OH)$_2$ and VR-MgO, in addition to aromatic carbons, also possess carbonate forms of carbon; see Figure 4. It is noteworthy to mention that the main peak for all the samples at ~284.8 eV corresponds to graphitic carbon groups. The high content of oxygen, the appearance of carbonates, and the absence of aliphatic carbons in VR-Ca(OH)$_2$ and VR-MgO suggest extensive oxidation of the asphalt carbons by the Ca(OH)$_2$ and MgO as opposed to KOH. Due to their high thermal stability, the Ca- and Mg- carbonates are still present even after the activation step at 400 °C. The details of the fitting results concerning the corresponding carbon functional groups are shown in Figure 4.

**Table 2.** Elemental composition of VR-KOH, VR-Ca(OH)$_2$, and VR-MgO estimated from XPS data.

| Sample | C% | O% | N% | S% | Ca% | Mg% |
|---|---|---|---|---|---|---|
| VR-KOH | 84.23 | 11.85 | 1.83 | 2.09 | - | - |
| VR-Ca(OH)$_2$ | 42.75 | 39.09 | 1.32 | - | 16.85 | - |
| VR-MgO | 34.09 | 52.69 | - | - | - | 13.22 |

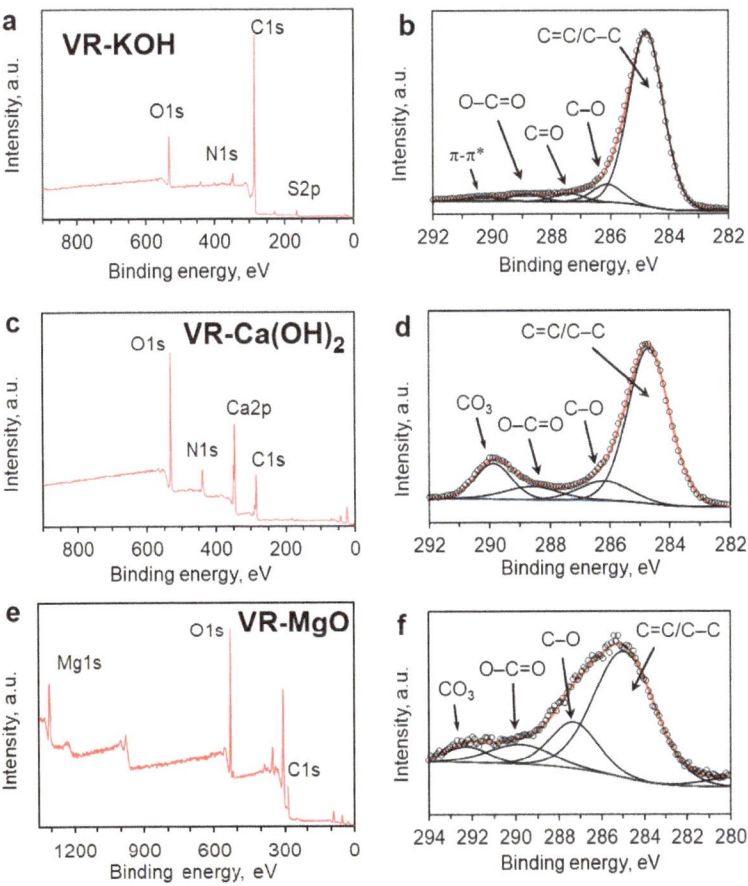

**Figure 4.** XPS survey spectra for (**a**) VR-KOH, (**c**) VR-Ca(OH)$_2$, and (**e**) VR-MgO. XPS high-resolution C 1s spectra for (**b**) VR-KOH, (**d**) VR-Ca(OH)$_2$, and (**f**) VR-MgO.

XPS, TGA, and N$_2$-adsorption isotherm analysis point out that the extent of the carbonaceous oxidation and the resulting microporosity depends on activating agents. KOH is the most efficient in generating micropores, while Ca(OH)$_2$ seems to extensively oxidize the carbonaceous materials in asphalt without micropores. It is also plausible to consider that the micropores were generated at an early stage in Ca(OH)$_2$ and MgO-activated samples. As such, with progressive oxidation in the activation step, the micropores enlarge and merge into meso- and macropores. Further study is required to elucidate the pore-formation steps and to find optimum conditions using the chemical agents or their combination. In addition, owing to the higher microporosity, VR-KOH was further analyzed using Raman spectroscopy as an indication of the degree of graphitization after the low-temperature activation step. The Raman spectra shown in Figure 5 compares VR-KOH as the sample with the highest carbon yield and the most porous with graphite powder. As can be seen in the figure, the VR-KOH sample exhibits a broad and pronounced D-band centered at ~1350 cm$^{-1}$. Furthermore, the intensity ratio of the D-band with the graphitic signature band (G-band) at ~1580 cm$^{-1}$ is much larger for VR-KOH than the graphite powder sample, which indicates a defective carbon material and a significant amount of amorphous carbon present in VR-KOH.

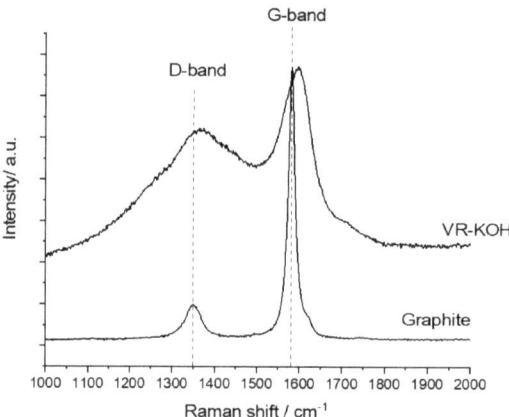

**Figure 5.** Raman spectra for VR-KOH and graphite powder.

Ongoing research in EDLC focuses mainly on increasing the capacitance by tuning the porosity of the carbon materials and utilizing novel nonaqueous electrolytes to increase the operating voltage of the supercapacitor [10,13,24,25]. It has been reported that the capacitance of asphalt-derived carbon materials demonstrates 127 to 140 F/g for discharge currents of 0.5 to 10 A/g using aqueous KOH electrolytes [26]. This is a promising result for using low-cost starting materials such as asphalts for energy-storage applications. However, the processes used in that particular work are multistep and tedious. It is of our interest herein to use simple straightforward chemical processes to modify the porosity, defect density, conductivity, and architecture of asphalt-based materials for symmetric supercapacitor applications.

Two main issues with asphalt-derived carbons related to supercapacitor applications are pore size distribution and electrical conductivity. The carbonization and activation processes performed on asphalts typically resulted in microporosity that significantly increases the BET surface area. However, as the supercapacitor energy storage is usually performed in liquid electrolyte media, access to the micropores is often hindered, especially when using high discharge currents. Furthermore, high annealing temperature (>800 °C) is usually employed to improve the electrical conductivity. Typically, when the annealing temperature is too high, the micropores collapse, dropping the surface area significantly. Due to the target activation temperature of 400 °C, the main focus of this work was on the effect of microporous structure and the effect of activating agents; in particular, $Ca(OH)_2$ and MgO as the main components of industrial waste. VR-KOH revealed the highest surface area with a predominantly microporous structure. Three-electrode setups with $KNO_3$ as an electrolyte were used to test the performance of the electrochemical supercapacitors, although due to the higher ionic conductivity of hydroxyl ($OH^-$) ions, the KOH electrolyte is expected to give a better capacitance than $KNO_3$. However, a neutral electrolyte such as $KNO_3$ is environmentally benign, which is in line with our objective of sustainability to utilize petrochemical waste. In addition, neutral electrolytes have been shown to expand the electrochemical window for carbon-based supercapacitors [27,28]. Specific capacitance (Csp) of the electrodes derived from VR-activated carbons was estimated using the expression shown below:

$$Csp = I/\{V \times (dv/dt) \times m\}$$

where $I$ is the CV curve area, $dv/dt$ is the scan rate, $V$ is the potential range, and $m$ is the mass of the active carbon materials. Supercapacitors require an accessible surface area; therefore, the high specific capacitance of 91.91 F/g was achieved for VR-KOH. On the other hand, carbon black with a specific surface area of 266 $m^2$/g shows a specific capacitance of 48.21 F/g; see Figure 6 and Table 3. The rate performance data of different carbon electrodes in a 2-electrode cell configuration is shown in Figure 6c and was collected from CV performance at different scan rates. For all the samples, the trend shows a decreasing capacitance with increasing scan rates. The result is in agreement with Table 3, in which the performance of VR-Ca(OH)$_2$ is comparable to the carbon black reference sample. Furthermore, VR-KOH and VR-MgO, two samples with a large difference in microporous areas, show a comparable specific capacitance, especially at higher scan rates. The potential range showed a minimum non-faradaic current according to a 3-electrode configuration test for each sample, which is indicative of the cyclability of the activated VR electrodes. Thus, it is not expected for any significant decay in capacity due to the electrochemical side reaction of the electrode and electrolyte. Nevertheless, capacity loss due to the mechanical detachment of the carbon powders is possible. Systematic investigation with XPS and SEM-EDX of the electrodes before and after cycling will give a better picture of the long-term electrochemical stability of the activated carbons. However, this is not the focus of our manuscript, which mainly highlights activating petroleum waste and its potential application. As an indication of the short-term cycle stability of VR-KOH, we provided cyclic voltammetry data of the VR-KOH electrode in a symmetric 2-electrode configuration showing similar current values for cycles 1 and 15 (Figure 6d).

Table 3. Summary of specific capacitance analyzed from 2-electrode cell measurements.

| Sample | Device Capacitance * (F), $\times 10^{-3}$ | Electrode-Specific Capacitance (F), $\times 10^{-2}$ | Powder-Specific Capacitance (F/g) | Carbon-Specific Capacitance (F/g) | Carbon Content%, from XPS Data |
|---|---|---|---|---|---|
| VR-KOH | 6.00 | 2.40 | 91.91 | 108.13 | 85.00 |
| VR-Ca(OH)$_2$ | 5.10 | 2.04 | 17.06 | 39.67 | 43.00 |
| VR-MgO | 4.10 | 1.64 | 27.39 | 74.01 | 37.00 |
| Carbon black | 8.20 | 3.28 | 48.21 | 48.21 | 100.00 |

* Values were calculated from the slopes of the scan rate versus the average current plot. Repeated CV measurements show about a 4% of the standard deviation of current values at 100 mV/sec and 0.4 V.

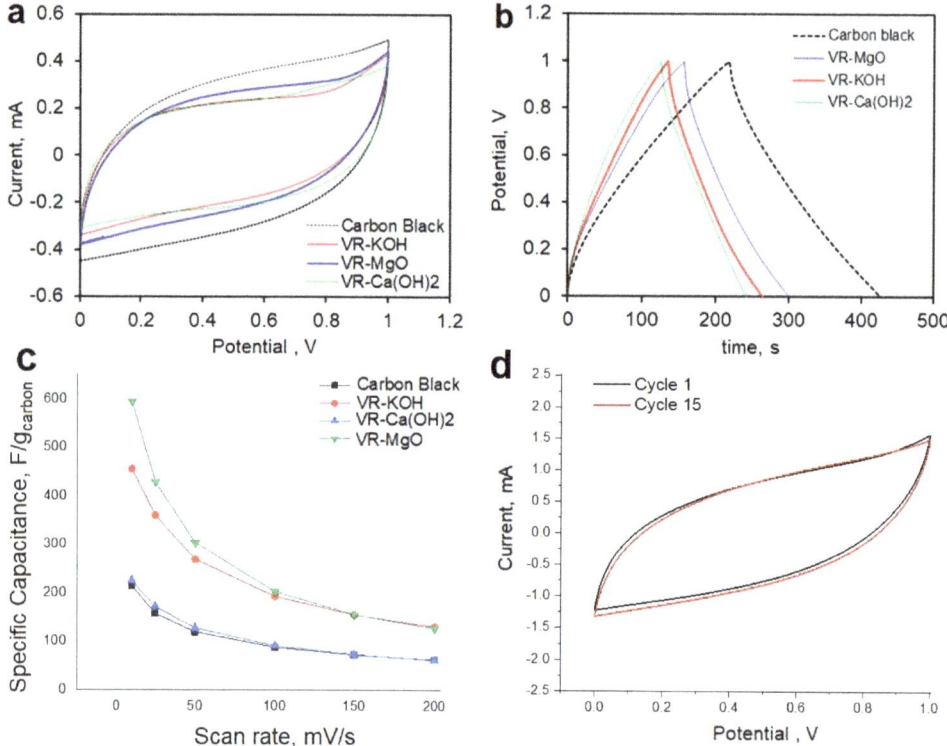

**Figure 6.** (**a**) Cyclic voltammetry of carbon black, VR-KOH, VR-Ca(OH)$_2$, and VR-MgO at 10 mV/s scan rate; (**b**) charge–discharge curves at a current density of 0.200 mA in 1 M KNO$_3$ electrolyte; (**c**) variation of specific capacitance with a scan rate of carbon electrode performed in 2-electrode cell configurations in 1 M KNO$_3$ electrolyte; and (**d**) cyclability of VR-KOH electrode in 1 M KNO$_3$ electrolyte.

The fact that both VR-Ca(OH)$_2$ and VR-MgO exhibit similar capacitance with carbon black, and that VR-KOH shows only about a 50% increase in capacitance indicates that the capacitance mainly comes from macropores. So, the micropores in KOH were not fully utilized. This could be due to the nature of the ink binder used for electrode preparation, the cell assembly, or it just being the intrinsic nature of the micropores that are too small for the solvated ions in the electrolyte. Further details of the importance of the electrode preparation are beyond the scope of the work and will be addressed in future reports.

## 4. Conclusions

In summary, vacuum residue directly acquired from the extraction facilities of the local industry was used to prepare activated carbons using various activating agents, such as KOH, Ca(OH)$_2$, and MgO. Characterizations of the final activated carbons derived from the vacuum residue resulted in the highest porosity for VR-KOH. Ca(OH)$_2$ and MgO as activating agents might require higher temperatures. However, the electrochemical supercapacitor performances of VR-Ca(OH)$_2$ and VR-MgO reveal being comparable with carbon black, which is typically made at >900 °C of the activation temperature. When the powder capacitance values are normalized with their corresponding BET surface area values, VR-KOH ranks the lowest, while VR-MgO and VR-Ca(OH)$_2$ exhibit higher numbers similar to the nonporous carbon black. This comparison highlights the discrepancies of surfaces in micropores for the capacitive current generation. This indicates that the micropores in VR-KOH are too narrow for the hydrated K$^+$ and NO$_3^-$ species. Future work to optimize the activation step to shift the porosity distribution in VR-KOH to larger diameters is expected to enhance the specific capacity of the activated carbon sample.

**Author Contributions:** Conceptualization, A.R. and A.S.J.; methodology, A.R. and A.S.J.; validation, A.A. (Abdualilah Albaiz), M.A., A.A. (Abdullah Alzahrani), A.R. and A.S.J.; formal analysis, A.A. (Abdualilah Albaiz), M.A. and A.A. (Abdullah Alzahrani); investigation, A.A. (Abdualilah Albaiz), M.A. and A.A. (Abdullah Alzahrani); resources, A.R. and A.S.J.; data curation, A.A. (Abdualilah Albaiz), M.A., A.A. (Abdullah Alzahrani), H.A., A.R. and A.S.J.; writing—original draft preparation, A.A. (Abdualilah Albaiz), M.A., A.A. (Abdullah Alzahrani), A.R. and A.S.J.; writing—review and editing, A.R. and A.S.J.; visualization, A.R. and A.S.J.; supervision, A.R. and A.S.J.; project administration, A.R. and A.S.J.; funding acquisition, A.S.J. All authors have read and agreed to the published version of the manuscript.

**Funding:** King Fahd University of Petroleum and Minerals (KFUPM), the Deanship of Research Oversight, and Coordination funding project DF191019.

**Data Availability Statement:** The data is available upon request from the corresponding authors.

**Acknowledgments:** The authors thank the financial support of King Fahd University of Petroleum and Minerals (KFUPM), the Deanship of Research Oversight, and Coordination for funding this work through project DF191019.

**Conflicts of Interest:** The authors declare no conflict of interest.

## References

1. Simon, P.; Gogotsi, Y. Materials for electrochemical capacitors. *Nat. Mater.* **2008**, *7*, 845–854. [CrossRef] [PubMed]
2. Miller, J.R.; Simon, P. Materials science-electrochemical capacitors for energy management. *Science* **2008**, *321*, 651–652. [CrossRef] [PubMed]
3. Zhang, L.L.; Zhao, X.S. Carbon-based materials as supercapacitor electrodes. *Chem. Soc. Rev.* **2009**, *38*, 2520–2531. [CrossRef] [PubMed]
4. Jiang, H.; Yang, L.; Li, C.; Yan, C.; Lee, P.S.; Ma, J. High-rate electrochemical capacitors from highly graphitic carbon-tipped manganese oxide/mesoporous carbon/manganese oxide hybrid nanowires. *Energy Environ. Sci.* **2011**, *4*, 1813–1819. [CrossRef]
5. Frackowiak, E.; Béguin, F. Carbon materials for the electrochemical storage of energy in capacitors. *Carbon* **2001**, *39*, 937–950. [CrossRef]
6. Pandolfo, A.G.; Hollenkamp, A.F. Carbon properties and their role in supercapacitors. *J. Power Sources* **2006**, *157*, 11–27. [CrossRef]
7. Béguin, F.; Presser, V.; Balducci, A.; Frackowiak, E. Carbons and Electrolytes for Advanced Supercapacitors. *Adv. Mater.* **2014**, *24*, 2219–2251. [CrossRef]
8. Frackowiak, E. Carbon materials for supercapacitor application. *Phys. Chem. Chem. Phys.* **2007**, *9*, 1774–1785. [CrossRef]
9. Shi, H. Activated carbons and double layer capacitance. *Electrochim. Acta* **1996**, *41*, 1633–1639. [CrossRef]
10. Li, Y.; Van Zijll, M.; Chiang, S.; Pan, N. KOH modified graphene nanosheets for supercapacitor electrodes. *J. Power Sources* **2011**, *196*, 6003–6006. [CrossRef]
11. Nasrin, K.; Sudharshan, V.; Subramani, K.; Sathish, M. Insights into 2D/2D MXene Heterostructures for Improved Synergy in Structure toward Next-Generation Supercapacitors: A Review. *Adv. Func. Mater.* **2022**, *32*, 2110267. [CrossRef]
12. Angizi, S.; Alem, S.A.A.; Pakdel, A. Towards Integration of Two-Dimensional Hexagonal Boron Nitride (2D h-BN) in Energy Conversion and Storage Devices. *Energies* **2022**, *15*, 1162. [CrossRef]
13. Yao, Y.; Ma, C.; Wang, J.; Qiao, W.; Ling, L.; Long, D. Rational Design of High-Surface-Area Carbon Nanotube/Microporous Carbon Core–Shell Nanocomposites for Supercapacitor Electrodes. *ACS Appl. Mater. Interfaces* **2015**, *7*, 4817–4825. [CrossRef]
14. Belaineha, D.; Brooke, R.; Sania, N.; Say, M.G.; Håkansson, K.M.O.; Engquist, I.; Berggren, M.; Edberg, J. Printable carbon-based supercapacitors reinforced with cellulose and conductive polymers. *J. Energy Storage* **2022**, *50*, 104224. [CrossRef]
15. Zhao, S.; Zheng, K.; Zhang, Z.; Wang, H.; Ren, J.; Li, H.; Jiang, F.; Liu, Y.; Cao, H.; Fang, Z.; et al. A facile synthesis strategy of fungi-derived porous carbon-based iron oxides composite for asymmetric supercapacitors. *Ceram. Int.* **2022**, *48*, 9197–9204. [CrossRef]
16. Wang, T.; Salvatierra, R.V.; Jalilov, A.S.; Tian, J.; Tour, J.M. Ultrafast Charging High Capacity Asphalt-Lithium Metal Batteries. *ACS Nano* **2017**, *11*, 10761–10767. [CrossRef]
17. Lee, K.S.; Parl, M.; Choi, S.; Kim, J.-D. Preparation and characterizaton of of N, S-codoped activated carbon-derived asphaltene used as electrode material for an electric double layer capacitor. *Colloids Surf. A* **2017**, *529*, 107–112. [CrossRef]
18. Enayat, S.; Tran, M.K.; Salpekar, D.; Kabbani, M.A.; Babu, G.; Ajayan, P.M.; Vargas, F.M. From crude oil production nuisance to promising energy storage material: Development of high-performance asphaltene-derived supercapacitors. *Fuel* **2020**, *263*, 116641. [CrossRef]
19. Jalilov, A.S.; Li, Y.; Tian, J.; Tour, J.M. Ultra-High Surface Area Activated Porous Asphalt for $CO_2$ Capture through Competitive Adsorption at High Pressures. *Adv. Energy Mater.* **2017**, *7*, 1600693. [CrossRef]

20. Jalilov, A.S.; Ruan, G.; Hwang, C.C.; Schipper, D.E.; Tour, J.J.; Li, Y.; Fei, H.; Samuel, E.L.; Tour, J.M. Asphalt-Derived High Surface Area Activated Porous Carbons for Carbon Dioxide Capture. *ACS Appl. Mater. Interfaces* **2015**, *7*, 1376–1382. [CrossRef]
21. Assen, N.v.d.; Jung, J.; Bardow, A. Life-cycle assessment of carbon dioxide capture and utilization: Avoiding the pitfalls. *Energy Environ. Sci.* **2013**, *6*, 2721–2735. [CrossRef]
22. Wang, W.; Liu, X.; Wang, P.; Zheng, Y.; Wang, M. Enhancement of $CO_2$ mineralization in $Ca^{2+}$-/$Mg^{2+}$-rich aqueous solutions using insoluble amine. *Ind. Eng. Chem. Res.* **2013**, *52*, 8028–8033. [CrossRef]
23. Grasa, G.; Abanades, J.C.; Anthony, E.J. Effect of partial carbonation on the cyclic CaO carbonation reaction. *Ind. Eng. Chem. Res.* **2009**, *48*, 9090–9096. [CrossRef]
24. Ko, T.-H.; Hung, K.-H.; Tzeng, S.-S.; Shen, J.-W.; Hung, C.-H. Carbon nanofibers grown on activated carbon fiber fabrics as electrode of supercapacitors. *Phys. Scr.* **2007**, *T129*, 80–84. [CrossRef]
25. Dolah, B.N.M.; Othman, M.A.R.; Deraman, M.; Basri, N.H.; Farma, R.; Talib, I.A.; Ishak, M.M. Supercapacitor Electrodes from Activated Carbon Monoliths and Carbon Nanotubes. *J. Phys. Conf. Ser.* **2013**, *431*, 012015. [CrossRef]
26. Song, X.Y.; Ma, X.L.; Yu, Z.Q.; Ning, G.Q.; Li, Y.; Sun, Y.Z. Asphalt-derived hierarchically porous carbon with superior electrode properties for capacitive storage devices. *ChemElectroChem* **2018**, *5*, 1474–1483. [CrossRef]
27. Bichat, M.P.; Raymundo-Piñero, E.; Béguin, F. High voltage supercapacitor built with seaweed carbons in neutral aqueous electrolyte. *Carbon* **2010**, *48*, 4351–4361. [CrossRef]
28. Chen, D.; Li, Z.; Jiang, J.; Wu, J.; Shu, N.; Zhang, X. Influence of electrolyte ions on rechargeable supercapacitor for high value-added conversion of low-grade waste heat. *J. Power Sources* **2020**, *465*, 228263. [CrossRef]

**Disclaimer/Publisher's Note:** The statements, opinions and data contained in all publications are solely those of the individual author(s) and contributor(s) and not of MDPI and/or the editor(s). MDPI and/or the editor(s) disclaim responsibility for any injury to people or property resulting from any ideas, methods, instructions or products referred to in the content.

Article

# Valorization of Exhausted Olive Pomace for the Production of a Fuel for Direct Carbon Fuel Cell

Najla Grioui [1,2,*], Amal Elleuch [1,2], Kamel Halouani [1,2] and Yongdan Li [3,4,5,6]

1. Laboratory of Systems Integration and Emerging Energies (LR21ES14), National Engineering School of Sfax (ENIS), University of Sfax, IPEIS Road Menzel Chaker km 0.5, P.O. Box 1172, Sfax 3018, Tunisia
2. Digital Research Center of Sfax, Technopole of Sfax, Sakiet Ezzit, P.O. Box 275, Sfax 3021, Tunisia
3. State Key Laboratory of Chemical Engineering, Tianjin University, Tianjin 300072, China
4. Tianjin Key Laboratory of Applied Catalysis Science and Technology, School of Chemical Engineering, Tianjin University, Tianjin 300072, China
5. Collaborative Innovation Center of Chemical Science and Engineering Tianjin, Tianjin 300072, China
6. Department of Chemical and Metallurgical Engineering, Aalto University, FI-00076 Aalto, Finland
* Correspondence: najla.grioui@ipeis.usf.tn; Tel.: +216-96766786

**Abstract:** In this study, exhausted olive pomace (EOP) biochar prepared by carbonization at 400 °C is investigated as a fuel in a direct carbon fuel cell (DCFC) with an electrolyte-supported configuration. The feasibility of using the EOP biochar in the DCFC is confirmed, showing a maximum power density of 10 mW·cm$^{-2}$ at 700 °C. This limited DCFC performance is compared with other biochars prepared under similar conditions and interrelated with various biochar physico-chemical characteristics, as well as their impact on the DCFC's chemical and electrochemical reaction mechanisms. A high ash content (21.55%) and a low volatile matter (40.62%) content of the EOP biochar are among the main causes of the DCFC's limited output. Silica is the major impurity in the EOP biochar ash, which explains the limited cell performance as it causes low reactivity and limited electrical conductivity because of its non-crystal structure. The relatively poor DCFC performance when fueled by the EOP biochar can be overcome by further pre- and post-treatment of this renewable fuel.

**Keywords:** DCFC; exhausted olive pomace; biochar; electrochemistry; limitations

## 1. Introduction

The use of biomass and waste as fuel represents not only a feasible but also promising alternative to conventional fossil fuels. Electricity from biomass and waste has received attention in recent years. The main advantage of biomass energy-based electricity is that renewable fuel is often a byproduct, a residue, or a waste produced from different sources, such as wood residues, agricultural residues, or agro-industrial wastes and byproducts. In this context, the national Waste Management Agency estimates that Tunisia produces around six million tons of organic wastes per year, where 73% comes from household wastes, farms, and the agro-food industry, while 17% is due to olive oil pressing residues. Tunisia, as one of the world's top five producers of olive oil, collects a huge amount (around 300.000 tons/year) of exhausted olive pomace (EOP) residues yearly. EOP is one of the main byproducts obtained from olive oil extraction, which represents an important economic sector in Mediterranean countries and specifically in Tunisia. This agro-industrial waste has a variable composition, depending on the olive variety and the olive oil processing methods. EOP consists of a lignocellulosic matrix with polyphenolic compounds, uronic acids, oily residues, water-soluble fats, proteins, water-soluble carbohydrates, and water-soluble phenolic substances [1,2], and it is rich in potassium and poor in phosphorus and micronutrients [2]. Nevertheless, the disposal of EOP is one of the major environmental problems due to its phytotoxicity and antimicrobial properties [3]. It also increases the hydrophobicity and the infiltration rate of soil and decreases the water retention rate [4,5].

However, thanks to its lignocellulosic structure, EOP can be directly combusted in boilers and furnaces [6–8], generating pollutant emissions and bad smells. Other applications for this material have been developed in recent years, such as the recovery of high-added-value compounds, including oligosaccharides, sugars (d-glucose and d-xylose), phenols compounds (antioxidants), bioethanol, xylitol, and furfural [9,10]. To date, thermochemical processes, such as pyrolysis, gasification, and hydrothermal carbonization (HTC), are in progress to eliminate the problem of uncontrolled emissions and to recover energy from EOP [11–15]. At a moderately high temperature (400–500 °C) under an inert atmosphere, carbonization (slow pyrolysis) appears as a promising way to convert EOP, mainly to biochar. Furthermore, recent experimentation of biochar in a DCFC system has proven to be a very attractive and eco-friendly technique for power generation, which has gained attention from scientists and engineers. Indeed, several carbon materials emitted from biomass have been examined as fuels in DCFC, which basically converts, at a high temperature (600–900 °C), the chemical energy stored in these biomass-based carbon materials to electricity through electrochemical oxidation reactions without a gasification or combustion process [16,17]. Compared to other fuel cell types, a DCFC system has a higher attainable efficiency (80%) for power generation and a lower emission of carbon dioxide per unit of the generated electricity [18]. In addition, such a system has an entropy term ($\Delta S$) close to zero, explaining the 100% theoretical efficiency of the system [19]. DCFC can be categorized into three types according to the types of electrolyte materials employed: molten carbonates, molten hydroxides, and solid oxides. All of them require a high temperature (>800 °C) to achieve a promising DCFC performance. A DCFC based on solid oxide electrolyte is similar to a SOFC, but it is fed by solid carbon instead of hydrogen. Several studies have been conducted to analyze SOFCs' short-term and even long-term durability, while few studies have been conducted to analyze DCFCs' durability, with some studies being in an early stage. However, SOFCs suffer from poor robustness and durability in the presence of impurities, such as sulfur and chlorine [20,21]. DCFCs need to be durable in the presence of such impurities to effectively utilize solid carbon resources (such as coal, biomass, biochar, and wastes) as fuel. A test to achieve the maximum long-term durability for a DCFC was developed by Jiang et al. [22]. The cell was tested for 100 h in their study. It is noticeable that durability analyses of fuel cells have an important impact on the future of fuel cell technologies in terms of performance optimization and further development, commercialization, and deployment [23].

Several research works [24–26] tried to decrease the operating temperatures of DCFCs and SOFCs by using thinner electrolyte layers or employing low-temperature, solid ion-conducting materials. In this sense, a solid composite electrolyte composed of a molten salt phase, such as mixed carbonates, and an oxygen ion-conducting porous solid phase, such as samarium doped ceria (SDC), has been employed firstly for SOFCs and then adopted for DCFCs. The conductivity of this kind of composite electrolytes in an intermediate temperature (IT) range of 400 to 700 °C is around $10^{-2}$ to 1 S m$^{-1}$ much higher than a conventional solid electrolyte in a solid oxide fuel cell and a molten electrolyte in a molten carbonate fuel cell (MCFC) [27]. The high conductivity of such a composite electrolyte has been explained by the enhancing effect of the co-ionic conduction in the two phases. SOFCs based on a composite electrolyte have shown remarkably good cell performance. Similarly, the operation of a DCFC at an intermediate temperature range (600–750 °C) has been investigated in order to overcome the problem of high temperatures [28].

One potential alternative electrolyte consisting of a mixture of samarium-doped ceria (SDC) and molten carbonate ($Li_2CO_3/Na_2CO_3$ in a mole ratio of 2:1) has received more attention, thanks to its double ionic conduction ability toward carbonate ($CO_3^{2-}$) and oxide ($O^{2-}$) ions [17,19,28,29].

Anode reactions:

$$C + 2O^{2-} \rightarrow CO_2 + 4e^- \tag{1}$$

$$C + 2CO_3^{2-} \rightarrow 3CO_2 + 4e^- \tag{2}$$

Cathode reactions:

$$O_2 + 4\,e^- \to 2O^{2-} \tag{3}$$

$$O_2 + 2\,CO_2 + 4e^- \to 2CO_3^{2-} \tag{4}$$

When a DCFC is fueled by a biochar composed mainly of carbon, hydrogen, and oxygen ($C_xH_yO_z$), other side reactions can take place within the anode. Indeed, it appears that the use of biochar as a direct fuel for DCFC systems demonstrates a peculiar behavior, notably within the anodic active electrochemical zone, regarding its heterogeneous elemental composition (CHO contents), and the presence of surface functional groups moves the anodic reaction mechanisms from a theoretically complete carbon oxidation reaction to a series of inter-related light gaseous and carbon oxidation reactions, apart from several side chemical reactions.

Effectively, through the biochar skeleton, various volatiles and light gases (CO, $CO_2$, $H_2$, and $CH_4$) may be generated at a high temperature. The formed $CO_2$ and $H_2O$ at the anodic active electrochemical reaction sites (AERS) can be directly used as a gasifying agent within the anode compartment and can further chemically react with solid carbon toward CO and $H_2$ formation [29]:

Boudouard reaction:

$$2C + CO_2 \to CO \tag{5}$$

Water shift reaction:

$$C + H_2O \to CO + H_2 \tag{6}$$

The latter reactions are strongly favored at a high temperature and present a key role in DCFC performance as their gaseous products (CO and $H_2$) can easily diffuse and reach the reaction sites much more rapidly than solid carbon materials, contributing largely to power generation through the following reactions:

$$H_2 + CO_3^{2-} \to CO_2 + H_2O + 2e^- \tag{7}$$

$$H_2 + O^{2-} \to H_2O + 2e^- \tag{8}$$

$$CO + O^{2-} \to CO_2 + 2e^- \tag{9}$$

$$CO + CO_3^{2-} \to 2CO_2 + 2e^- \tag{10}$$

Based on these series of electrochemical reactions, the overall DCFC efficiency can be attributed mainly to CO and $H_2$-AERS interactions, rather than to the extremely limited solid carbon-AERS contact [19,29–31].

Practically all the experimental and numerical studies on DCFC anode kinetic mechanisms affirmed that the hypothesized 4-electron carbon electrochemical oxidation reaction (Equation (2)) is not sufficient to explain the recorded DCFC performance. The occurrence of a 2-electron CO oxidation (Equation (9)) and chemical Boudouard reaction (Equation (5)) is often illustrated. This chemical reaction is temperature and $CO/CO_2$ content dependent. The reverse Boudouard reaction is known to be fast at 700 °C. However, in a molten carbonate medium, it exhibits a strange behavior. Its rate may be much slower. Meanwhile, the 2-electron CO oxidation (Equation (9)), which is a consequence of the occurrence of the reverse Boudouard reaction, is known to have a limited kinetic rate below 650 °C and a significantly accelerated rate starting from 700 °C. The backward sense of the latter electrochemical reaction can also possibly occur, increasing the CO concentration within the anode. In this sense, Chen et al. [32] developed a macro-homogeneous model to assess the kinetics of the three aforementioned reactions and their dependence on several properties, such as anodic bed thickness and carbon conductivity. They concluded that the electrochemical mechanism is approximately three times as fast as the chemical reverse Boudouard reaction near the anodic current collectors.

Various research studies have focused on the investigation of biomass-based carbon materials' potential as fuels in DCFCs to reveal their efficacy as energy carriers. It has been found that both the performance and lifetime of DCFCs are notably affected by the physico-chemical properties of biomass-based carbon fuels [18,31,33]. Actually, the recorded low performance of biochar-fueled DCFCs hinders their further development.

Wang et al. [34] assessed the potential of reed biochar as a fuel in a direct carbon fuel cell based on a SDC-carbonate composite electrolyte and achieved the best maximum power density of about 378 mW·cm$^{-2}$ at 750 °C to date. They obtained this promising performance after using a KCl-washing pre-treatment on the raw biomass before pyrolysis. The effect of KCl washing in raw reed increased the structural disorder degree of the biochar during the pyrolysis process, leading to a high oxidation activity of the reed biochar and, subsequently, a good DCFC performance. KCl is known as one of the chemical activating agents used in the preparation of activated carbons for energy storage applications, such as supercapacitors and batteries [35,36]. KOH, $H_2SO_4$, and $ZnCl_2$ are other types of agents that can contribute to the activation of biomass precursors. These chemical agents ensure an increase in total porosity and micropore development, as well as an increase in the yield of the activation process. Gómez et al. [37] proposed a reaction mechanism, which was validated by mass spectroscopy analysis and thermodynamic calculations, to carry out activation through the use of a mixture of KOH and KCl. They concluded that the role of KCl consists of arise in the solubilization of carbonates that precipitate in its absence, hence lowering the contact between the liquid KOH and the carbon particles. They affirmed that the use of KCl as an additive results in the synthesis of activated carbons with lower amounts of KOH, which are, therefore, more available to be produced at large scales. Jayakumar et al. [38] achieved 360 mW·cm$^{-2}$ at 700 °C with sugar char as a fuel when using a molten antimony anode. Cai et al. [39] used a biochar derived from orchid tree leaves as fuel. They showed that the high content of $CaCO_3$ in the leaf biochar catalyzes the reverse Boudouard reaction and enhances the performance of DCFCs. Hao et al. [40] also found that carbon from magazine waste paper contains a high amount of magnesium calcite, which improves the thermal reactivity of the carbon materials. Chien and Chuang [41] used coconut coke as a fuel for an anode-supported DCFC and recorded a maximum power density of about 80 mW·cm$^{-2}$ at 800 °C. They affirmed, through CO and $CO_2$ pulse transient studies, that the increased cell performance was attributed to an increasing extent of electrochemical oxidation of CO, a product of the Boudouard reaction. Hao et al. [42] tested a DCFC with bamboo carbon as fuel and recorded a maximum power density of 156 mW·cm$^{-2}$ at 650 °C. They concluded that the inherent impurities, such as calcite ($CaCO_3$) and kaolinite ($Al_2Si_2O_5(OH)_4$), in the biochar might favor its thermal gasification and resulted in the enhanced performance of the intermediate-temperature DCFC. Elleuch et al. investigated almond shell (AS) [19] and olive wood (OW) [29] biochars as fuels in DCFCs supported by a 0.65 mmthick $Ce_{0.8}Sm_{0.2}O_{1.9}$ (SDC)-carbonate composite electrolyte layer and recorded a maximum power density of 107 and 105 mW·cm$^{-2}$, respectively, at 700 °C. They claimed that the high concentration of oxygen-containing groups is the main reason for the higher performance recorded, when compared to activated carbon. They also concluded that alkali and alkaline-earth metal oxides, such as $K_2O$, $Fe_2O_3$, and CaO, worked as the active catalysts for the anodic reaction by decreasing the electrochemical activation polarization in the case of the AS biochar. It is known that gasification reactions take place below 800 °C [43], and the gasification of carbon fuel is the limiting factor of DCFC performance when operating at a higher temperature range. The kinetics of these reactions can be catalyzed using several catalysts. Meanwhile, it has become a consensus that the alkali metal, alkaline earth metal, and transition metal are effective catalysts for carbon gasification [44–46], which are widely used in coal gasification research to obtain a competitive reaction rate at a lower temperature.

Li et al. [43] tested Ni, K, and Ca catalysts and claimed that all of them are suitable gasification catalysts to accelerate the carbon gasification rate, reduce the reaction temperature of the DCFC, and, thus, improve the cell performance. Cui et al. [27] tested the effect of carbonate as a catalyst and affirmed that it can also play a catalytic role in carbon gasification reactions. Tang et al. [47] used gadolinium-doped ceria (GDC) mixed with silver as the anode to catalyze the electrochemical oxidation of CO, while a Fe-based catalyst was loaded on the carbon fuel to enhance the Boudouard reaction. They mentioned an enhancement of about 10 times higher than that of a cell without any catalyst. Recently, a strontium slag and its derived catalyst were successfully introduced with the carbon materials at the anode in order to enhance a DCFC's performance by promoting the reverse Boudouard reaction [48]. Yu et al. [49] used corncob biochar as fuel and a single cell similar to the one used by Elleuch et al. [19,29], but with adding a printed anode layer (a mixture of NiO and SDC). They showed a maximum power density of 185 mW·cm$^{-2}$ at 750 °C.

The main objective of the present research is to explore the electrochemical capability of EOP biochar as fuel in a direct carbon fuel cell (DCFC), using an experimental correlation between the DCFC's power output and the physico-chemical properties of the fuel, in order to determine the main EOP biochar limiting properties when it is used as a fuel for DCFCs. A series of physico-chemical analyses are investigated for this purpose. Furthermore, the DCFC's chemical/electrochemical mechanisms are predicted with respect to the EOP biochar's physico-chemical properties and compared to similar DCFC configurations, which have successfully operated with other biochar fuels when prepared under the same conditions [19,29].

## 2. Materials and Methods

### 2.1. Raw Material

The exhausted olive pomace (EOP) used in the present study was recuperated from an extraction factory at Sfax city in Tunisia. The tested EOP sample is a mixture of residues of olive stones and olive pulp extracted from both traditional olive pressing and 3-phase centrifugation systems, followed by solvent extraction using n-hexane for olive pomace oil recovery.

The EOP was sieved using a 10–20 mesh (2.00 ± 0.85 mm) particle size. The sample of EOP was shipped to Galbraith Analytical Laboratory, (Knoxville, TN, USA), where it was analyzed to determine its elemental compositions and to estimate the high and low heating values. For the determination of lignin, cellulose, and hemicellulose, the first removal of soluble extractives was performed according to the standard TAPPI T-264 cm-97 [50]. Then, lignin and cellulose were determined according to the standard TAPPI T 222om-83 and TAPPI T 203 os-74, respectively [50], and holocellulose was determined according to Browning [51]. Hemicellulose concentration was calculated as the difference between holocellulose and cellulose. Ash content was determined according to the ASTM D 482. A proximate analysis (moisture and volatile matter) was performed using a Thermogravimetric Analyzer TGA Labsys (Setaram Instruments, Paris, France). About 20 mg of the sample was loaded into a platinum pan and heated at 20 °C min$^{-1}$ until 600 °C under a nitrogen atmosphere (nitrogen flow rate of 40 mL·min$^{-1}$). The weight loss from 30 to 150 °C was due to the moisture content; from 150 to 600 °C, it was due to the volatile matter content. The residual mass was assigned to a mixture of fixed carbon and ash. Fixed carbon was determined by the difference. The analytical results are summarized in Table 1.

The high heating value (HHV) of the EOP biomass feedstock is determined using the modified Dulong's formula based on the following equation [52]:

$$\text{HHV}(\text{MJ.kg}^{-1}) = (33.5 \times C + 142.3 \times H - 15.04 \times O - 15 \times N)/100 \tag{11}$$

where C, H, O, and N are the carbon, hydrogen, oxygen, and nitrogen contents (wt%), respectively.

The low heating value (LHV) is calculated as a function of the HHV and the hydrogen content [52]:

$$\text{LHV (MJ.kg}^{-1}) = \text{HHV} - 2.442 \times 8.936 \times (H/100) \tag{12}$$

Table 1. Proximate and elemental analysis of the EOP raw material and its derived biochar.

| Analysis | Sample | EOP Raw Material | EOP Biochar |
|---|---|---|---|
| Proximate analysis (%) | Moisture | 7.31 | 1.38 |
| | Ash | 10.91 | 21.55 |
| | Fixed carbon ** | 25.28 | 36.45 |
| | Volatile matter | 56.5 | 40.62 |
| Ultimate analysis * (%) | C | 39.45 | 65.7 |
| | H | 5.58 | 3.4 |
| | N | 2.68 | 0.88 |
| | S | <0.8 | 0.08 |
| | O ** | 41.2 | 29.93 |
| Molar formula | | $CH_{1.69}O_{0.78}N_{0.058}$ | $CH_{0.62}O_{0.34}N_{0.01}$ |
| Lignocellulosic material composition | Hemicellulose (%) | 13 | |
| | Cellulose (%) | 27 | |
| | Lignin (%) | 49 | |
| High heating value (HHV. MJ kg$^{-1}$) | | 14.43 | 22.21 |
| Low heating value (LHV. MJ kg$^{-1}$) | | 13.21 | 21.47 |

* Dry basic. ** By difference.

## 2.2. Carbonization Pilot Plant Description

The EOP carbonization was carried out in a pilot plant installed at ENIS, which is illustrated in Figure 1. It includes a metallic pyrolysis chamber connected to a combustor of recycled carbonization volatiles that are transported by two insulated gas channels. The combustor is connected to a heat exchanger, which heats the carbonization chamber through hot combustion gases. The carbonization temperature is controlled by a thermocouple. The carbonization chamber was filled with about 50 kg of the EOP and then underwent a first pre-heating step using a gasoline burner in order to carry out the drying phase of the EOP biomass. From a temperature of about 400 °C, the EOP biomass was decomposed, and the carbonization volatile gases, which were generated from the carbonization chamber and circulated through the insulated gas channel, replaced the gasoline and ensured the auto-feeding of the flame in the combustor. The EOP biochar was recuperated at the end of the carbonization experiment within the carbonization chamber. This carbonization experimental procedure took around 4 h.

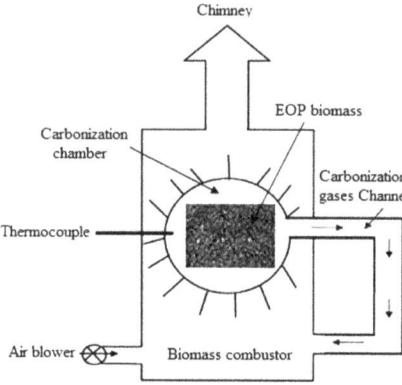

Figure 1. Schematic diagram of the EOP biomass carbonization plant.

## 2.3. Physico-Chemical Characterization of the EOP Biochar

Determining the elemental composition is a crucial analysis to be carried out for a biochar, which could be used, thereafter, as a fuel, to predict its heating capacity, combustion performance, and exhaust emissions. The percentages of C, H, N, and S (CHNS) of the biochar were determined using the Vario Micro Elementar CHNS system. The oxygen content was calculated by the difference.

The HHV of the biochar was calculated using Equation (11). The volatile matter, moisture, and ash content of the EOP biochar were determined using a thermogravimetric analysis and the ASTM D 482, respectively. XRF was carried out using the S4 Pioneer system in order to determine the chemical composition of the EOP biochar ash. X-ray diffraction (XRD) analysis was performed in order to identify the crystal structure of the EOP biochar, using a Philips X'pert Pro super Diffractometer with monochromatic Co K$\alpha$ radiation in the $2\theta$ range from $10°$ to $80°$ at a scanning rate of $2°\cdot min^{-1}$. The layer dimension perpendicular to the basal plane, $L_c$, of the EOP biochar was obtained from the (002) reflection angle following the Debye–Scherrer equation, whereas the d-spacing was determined according to the Bragg formula, with n being equal to 1 and $\theta$ representing the (002) reflection angle [18].

The porous morphology of a biochar is an important property to investigate, considering its relationship to the surface area, the distribution of the pore's diameter, its porosity, and the effect of these parameters on gaseous species absorption.

Representative SEM showing the primary particle sizes of the EOP biochar was recorded using a Hitachi S-4800 scanning electron microscope. In addition, an Autopore IV 9500 mercury porosimeter (Micrometrics, York, PA, USA) was used to measure the pore volume distribution, the surface area, and the porosity of the EOP biochar by measuring the amount of mercury penetrating into the sample pores [53].

The temperature-programmed oxidation analysis (TPO) of the EOP biochar sample was carried out using a thermobalance (NETZSCH STA, 449F3) with an air flow of about $80\ mL\cdot min^{-1}$ at the standard conditions of temperature and pressure (STP) and at a heating rate of $10\ °C\cdot min^{-1}$. The temperature ranged from room temperature to $1000\ °C$.

The EOP biochar's chemical functional groups were identified by FTIR analysis using a Nicolet 380 spectrometer (Thermo-Scientific, Waltham, MA, USA) in the range of $4000$–$400\ cm^{-1}$. The standard IR spectra of hydrocarbons were used to identify the functional groups of the biochar components.

## 2.4. Electrochemical Test of the EOP Biochar in Direct Carbon Fuel Cell

The direct carbon fuel cell (DCFC) used in this study was similar to the one employed in our previous works [19,29]. Samarium-doped ceria (SDC) combined with molten carbonate (MC) consisting of a $Li_2CO_3/Na_2CO_3$ (Chemical reagent, Tianjin, China) eutectic mixture in a mole ratio of 2:1 was used as the electrolyte material. The SDC sample ($Ce_{0.8}Sm_{0.2}O_{1.9}$) was synthesized via an oxalate co-precipitation process. All the chemicals were used as received without further purification. The stock solution was made by mixing and dissolving 60.3913 g of cerium nitrate hexahydrate ($Ce(NO_3)_3\cdot 6H_2O$, 99%, Chemical reagent, Tianjin, China) and samarium nitrate hexahydrate ($Sm(NO_3)_3\cdot 6H_2O$) in 100 mL of distilled water. A total of 6.0624 g of samarium oxide ($Sm_2O_3$, 99.5%, Chemical reagent, Tianjin, China) was dissolved in 10 mL of nitric acid (Chemical reagent, Tianjin, China) and 15 mL of distilled water to obtain samarium nitrate. Aqueous oxalate acid solutions ($H_2C_2O_4\cdot 2H_2O$, Chemical reagent, Tianjin, China) with a concentration of $0.1\ mol\cdot L^{-1}$ were used as the precipitant. In a typical synthetic procedure, 550 mL of distilled water mixed with the oxalate acid solution (Chemical reagent, Tianjin, China) was dripped at a speed of $3\ mL\ min^{-1}$ into the precipitant solution under 250 rpm vigorous stirring at room temperature to form a white precipitate. The resultant suspension, after homogenizing for 1 h, was filtered via suction filtration. The precipitate cake was washed repeatedly with distilled water and ethanol, followed by drying at $100\ °C$ for 24 h to obtain the SDC precursor. The obtained SDC precursor was sintered at $700\ °C$ for 2 h to form a pale yellow

SDC powder ($Ce_{0.8}Sm_{0.2}O_{1.9}$). The binary carbonate powder, i.e., $Li_2CO_3/Na_2CO_3$ in a mole ratio of 2:1, was prepared. The composite electrolyte material powder was obtained through mixing the two powders (carbonate powder and SDC powder) in a weight ratio of 3:7 by ball milling for 2 h.

The used composite cathode powder consisted of 30 wt% composite electrolyte and 70 wt% $Li_xNi_{1-x}O$ powders. The composite cathode powder was also prepared through 2 h ball mill mixing, sintering at 700 °C for 2 h, and grinding. The lithiated nickel oxide ($Li_xNi_{1-x}O$) was prepared through 2 h ball mill mixing of NiO with lithium hydroxide monohydrate ($LiOH·H_2O$, 90%, Chemical reagent, Tianjin, China) powders in a 1/1 mol% ratio, sintering at 700 °C for 2 h, and grinding. The fine NiO powder was produced through heating a proper amount of $Ni(NO_3)_2·6H_2O$ (99%, Chemical reagent, Tianjin, China) powder at 700 °C for 2 h until it combusted.

The SDC/MC powder (0.25 g) was first uniaxially pressed in a die at 1 MPa for 60 s to form yellow electrolyte discs. A total of 0.15 g of the composite cathode powder ($Li_xNi_{1-x}O$-SDC/MC) was added using a 60-mesh sieve to the electrolyte discs. Then, a single isostatic pressing at 500 MPa (30 kpsi) was performed for 30 s to form the DCFC cell pellet. The cell pellet was then sintered at 700 °C for 2 h in air. The sintered DCFC pellet had a diameter of 13 mm and was 1 mm thick. Silver paste was brush painted on both sides of the pellet for the current collection. The experimental protocol of the DCFC pellet preparation is described in Figure 2.

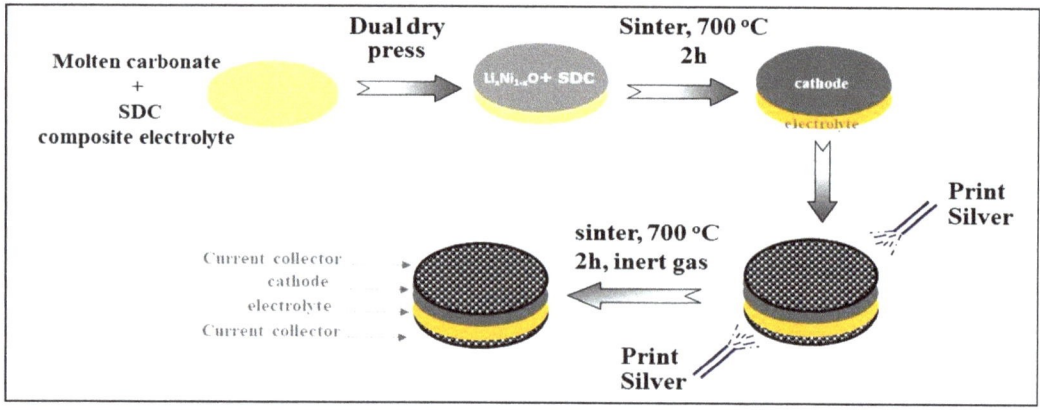

**Figure 2.** Experimental protocol of DCFC pellet preparation.

The obtained pellet was mounted in a stainlesssteel cell holder (Figure 3) serving for the current collection and for gas distribution. The DCFC pellet was placed with the cathode downwards and then sealed into a tubular furnace controlled by a K-type thermocouple.

Typical carbon material loading was 300 mg for the EOP biochar fuel. The obtained EOP biochar from carbonization was ground and sieved between 40 and 60 meshes to fit the application condition as a fuel in a DCFC system. A sealant was cured in situ during the heating of the cell to isolate the cathode and anode chambers. The fuel chamber was continuously purged under nitrogen at a 100 mL·min$^{-1}$ flow rate. The cell temperature was raised to 100 °C in an ambient air environment in order to ensure seal formation. Then, the inert gas purge was started, and the temperature was increased up to 700 °C. The cell was held at this temperature to assess the electrochemical performance of the DCFC fed by the EOP biochar. The cathode chamber was fed by a mixture of $O_2$ (60 mL·min$^{-1}$) and $CO_2$ (120 mL·min$^{-1}$). A Versastat 3 Potentiostat-Galvanostat equipped with the Versastudio software for automatic data collection was used to generate the DCFC cell polarization curves.

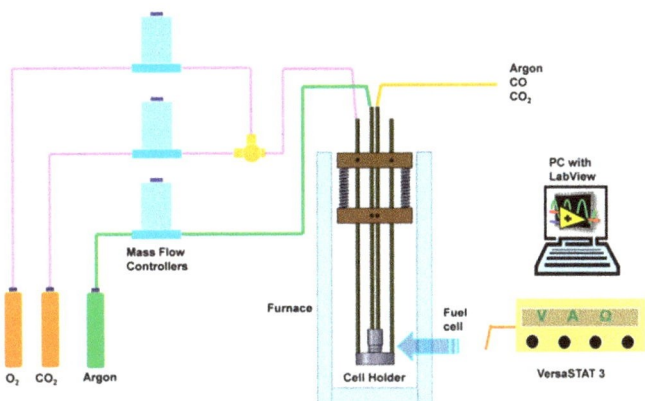

**Figure 3.** Operation mode of the DCFC.

## 3. Results and Discussions

### 3.1. Feedstock Characterization

The physico-chemical properties of the EOP raw materials are significantly important for the assessment of their byproduct potential and quality. Proximate and ultimate analyses are among the easiest ways to study the fuel characteristic of solid materials. The results of these analyses are presented in Table 1.

The EOP biomass feedstock is mainly composed of carbon (39.45%) and oxygen (41.2%), with a low percentage of nitrogen (2.68%) and nearly no sulfur content (<0.8%). The EOP biomass composition aligns with other solid wastes reported in the literature [53].

Based on the elemental composition results, the approximate molar chemical formula of the EOP biomass is $CH_{1.69}O_{0.78}N_{0.058}$.

The proximate analysis results show that the EOP contains 7.31%, 56.5%, 10.91%, and 25.28% of moisture, volatile matter, ash, and fixed carbon, respectively (Table 1). The values obtained from the proximate analysis are in the range reported for most agricultural and forest wastes [54,55].

The chemical analysis (Table 1) also shows that the EOP has higher lignin content (49%) than cellulose and hemicellulose contents (holocellulose) (40%). This finding is in line with that reported by Jauhiainen et al. in [5,9], which means that the EOP has a different nature then common biomasses and wastes.

The high heating value (HHV) of the EOP (14.43 $MJ \cdot kg^{-1}$) is also in the range of reported HHV for other biomass feedstocks (11–40 $MJ \cdot kg^{-1}$) cited in the literature and used as fuels [56].

### 3.2. EOP Biochar Characterization

In this section, the deep characterization of the physico-chemical and structural properties of the EOP biochar are presented, aiming to assess its possible application as a fuel in a DCFC system. The performance and the durability of DCFCs are dependent on biochar properties. Some properties appear as inhibitors, whereas others are favorable for the electrochemical conversion of biochar in DCFCs.

The ultimate analysis of the EOP biochar is also presented in Table 1 in comparison with the raw material. Carbon, as the main element in the obtained biochar (65.7%), is present in significantly high amount compared to the EOP biomass feedstock (39.45%). However, the oxygen (29.97%) content significantly decreases in comparison to the EOP raw material (41.2%). The low oxygen content in the EOP biochar sample is due to the dehydration and decarbonylation/decarboxylation reactions occurring during carbonization. The nitrogen and sulfur contents in the EOP biochar are low (0.89% and 0.08%, respectively) but still in an acceptable level for fuel application. Nitrogen and sulfur in the EOP biochar can be considered as impurities that may affect the DCFC performance negatively [57,58]. The atomic ratios of hydrogen to carbon (H/C) and oxygen to carbon (O/C) are considered to be important indicators of a fuel's quality. It is observed that the H/C and O/C ratios of the EOP biochar are lower than those of the raw EOP biomass. These lower values are due to the increase in carbon content and the decrease in oxygen and hydrogen contents during carbonization due to dehydrogenative polymerization and dehydrative polycondensation reactions [58]. In fact, the low H/C ratio (0.62) of the EOP biochar is indicated on its structural modification, which shifts to a higher content of aromatic compounds that are more resistant to thermal degradation and, thus, remain recalcitrant [59].

When neglecting the biochar's sulfur content, the approximate molar chemical formula of the biochar can be expressed as $CH_{0.62}O_{0.34}N_{0.01}$.

On the basis of the dry biochar elemental composition, the HHV is calculated using Equation (11) and is found to be equal to 22.21 MJ·kg$^{-1}$. The HHV of the EOP biochar is higher than the EOP biomass and most coals, confirming its potential use as an energy source.

The proximate analysis of the EOP biochar (Table 1) shows a high ash content (21%) and a low moisture content (1.38%). However, the volatile matter and fixed carbon contents of the EOP biochar are 46% and 31.62%, respectively. The EOP biochar exhibits a high content in ash (21.55%), compared to only 6.4% in the case of AS biochar [19] and 7.49% in the case of OW biochar [29]. The ash content increases after the carbonization of the EOP biomass (Table 1), which follows the same tendency of all carbonized biomasses. Considering the ash content, the literature has the unanimous idea that low ash content is desirable for biochar utilization as a fuel in DCFC systems. The mineral matter forming the ash reduces the lifetime of a DCFC by blocking the active reaction sites of the anode, but it is important to note that some mineral impurities, which derive from the biochar fuel source, are considered as inhibitors, while other minerals act as catalysts for the DCFC electrochemical and chemical reaction processes [19].

An analysis of ash composition is critically recommended in order to check if an EOP biochar falls under the IBI standards, which states that, "because of the known presence of heavy metals and organic pollutants in bio-solids, care must be taken during thermochemical conversion to avoid harmful air emissions as well as the accumulation of toxicants in the final carbonaceous skeleton of EOP biochar [60]". In addition, the composition of minerals is crucial to analyze before using a biochar as a fuel in a DCFC system, as previously revealed. Rady et al. [61] mentioned that $Al_2O_3$ and $SiO_2$ act as the main inhibitor for the DCFC electrochemistry. However, systematic studies on the effects of these species at various concentrations are required before they can be accurately categorized as either inhibitors or catalysts and to know the extent of their inhibitive or catalytic effects.

An investigation of the ash composition was carried out using XRF analysis; the results show that the EOP biochar contains a high amount of silica compared to the alkali and alkaline earth metals (Mg, Ca, K, . . . ) (Table 2), together representing more than 30% of the ash composition. Si is the dominant compound in the EOP biochar (55.21%), followed by Ca (16.3%) and K (13.72%), but Al is present with a low percentage fraction of about 1.79%.

The high content of $SiO_2$ may cause passivation in the DCFC and an instability in the electrochemical performance of the cell regarding its inhibitive character. This finding has been confirmed by Vutetakis et al. [62], who studied the effects of various mineral impurities on a fluidized bed DCFC and observed a sharp drop in current at high overpotentials, which was explained by the passivation of electrodes since a film was formed on their surfaces due to dissolved $Al_2O_3$, $SiO_2$, and $TiO_2$ from the coal ashes.

At a low gasification temperature (about 700 °C), it has been confirmed that a silica structure keeps its physical criteria but undergoes chemical changes through the presence of alkali metals [63]. In a biochar containing very high silica contents, such as rice husks and bagasse, alkali silicates may be formed through the reaction of silica with alkali metals. These resulting silicates are mesoporous with a limited surface area and can be induced in some other way to have a limited pore volume, as present in the residual biochar. It was also demonstrated that progressive heating of these silicates revealed the ability of trapping coke deposited within the pore media. As a result, the ash residuals showed significant organic contents, even after extensive additional oxidation in air [64].

Table 2. Mineral composition of the EOP biochar ashes obtained using XRF analysis.

| Non-Organicelemental Composition (%) | EOP Biochar |
|---|---|
| MgO | 0.96 |
| $Al_2O_3$ | 1.79 |
| $SiO_2$ | 55.21 |
| $P_2O_5$ | 3.33 |
| $SO_3$ | 2.06 |
| $K_2O$ | 13.72 |
| CaO | 16.31 |
| $Fe_2O_3$ | 3.93 |
| SrO | 0.40 |
| Cl | 1.98 |
| $TiO_2$ | 0.30 |

Furthermore, alkali and alkaline earth metals are identified as effective catalysts for some chemical reaction mechanisms in several applications, such as gasification [64], insitu catalytic fast pyrolysis [65], and even in a DCFC application.

Indeed, the presence of alkali and alkaline earth metals promotes the water–gas shift reaction under the gasification process and enhances the yield of $H_2$ and $CO_2$. Additionally, they not only boost the breakage and decarboxylation/decarbonylation reaction of the thermally labile hetero atoms of tar, but they also enhance the thermal decomposition of heavier aromatics. These impurities could also significantly enhance the decomposition of levoglucosan. It has been proven that alkaline earth metals show greater effect than alkali metals for these series of decomposition reactions, as reported previously.

Figure 4 presents the X-ray diffraction (XRD) patterns of the EOP biochar, which was performed to investigate its crystallographic structure and disorder.

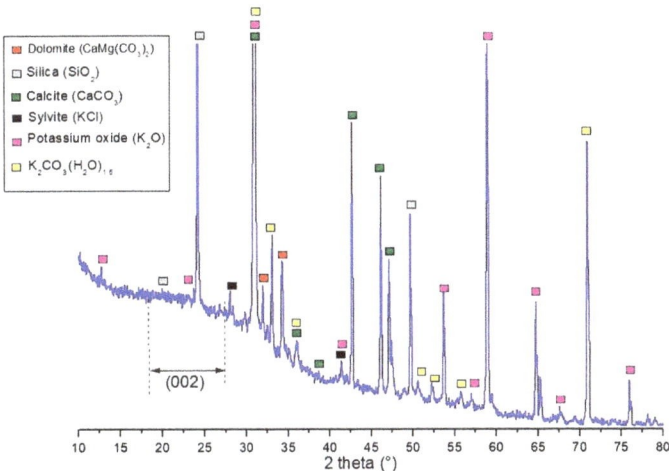

Figure 4. XRD patterns of the EOP biochar.

The pattern of the EOP is different from other biochars prepared under similar carbonization conditions [19,29] as it presents a series of sharp peaks at different diffraction angles, confirming the existence of high inorganic material content in the EOP biochar, including mainly silica ($SiO_2$), sylvite (KCl), calcite ($CaCO_3$), and dolomite ($CaMg(CO_3)_2$). The obtained XRD pattern of the EOP biochar sample has a similar trend to the XRD patterns reported by other researchers, who used a different feedstock, such as wood, grass, corn straw, peanut straw, and claimed that the high numbers of observed peaks is ascribed to various crystal components [66–68]. For example, peaks at $2\theta$ = 20.88 and 25° were designated as silica, while the peak at $2\theta$ = 50.18° was designated as feldspar in a raw silica sample [69]. Thus, a comparison of the EOP biochar pattern with the Powder Diffraction and Standards (PDF) was conducted to identify the corresponding peak of each mineral.

The XRD pattern of the EOP biochar indicates similarity in the broad peak corresponding to the (002) graphitic basal plane reflection of graphite. This peak exists at around $2\theta$ = 25°, slightly shifting from the peak location of graphite commonly found at 26.5°, thus indicating the amorphous and turbostatic structure of the EOP biochar. The shift was also reported by Konsolakis et al. [31] when investigating the XRD patterns of biochars from pistachio shells, pecan shells, and sawdust.

The XRD arrangement for AS biochar [19], OW biochar [29], and corn cob biochar [49], in comparison to graphite, also demonstrate one distinct peak at $2\theta$ = 29°, 26°, and 24°, respectively. This peak corresponds to the (002) reflection, and its presence has been revealed at comparatively lower values when compared to graphite, proving that the latter three biochar samples exhibit a crystallographically disordered structure similar to the EOP biochar investigated here.

In the case of the EOP biochar, this peak intensity is weak due to the overlap with the intensive peak of silica seen at the same diffraction angle range. This is expected due to the high content of silica present in the EOP biochar, as reported in Table 2. Meanwhile, this behavior was previously reported by Ahmad et al. [69] in their XRD analysis of date palm waste biochar and its derived composite made of the same biochar mixed with silica. The peak of (002) graphitic basal plane depicted in the case of the biochar pattern is totally removed in the case of the silica-composited biochar pattern.

The quantitative crystallite parameters of the EOP biochar, including the interplanar distance ($d_{002}$) and the stacking height ($L_c$), are also shown in Table 3. The obtained value proves the turbostatic structure of the EOP biochar as the $d_{002}$ value (0.3622 nm) is slightly higher than the graphite $d_{002}$ value (0.36 nm), and the $L_c$ (24.02 nm) is much higher than the $L_c$ (13.8 nm) value assigned to graphite.

**Table 3.** Summarized properties of the EOP biochar sample.

|  | EOP Biochar |
|---|---|
| Crystalline parameters |  |
| $d_{002}$ (nm) | 0.3622 |
| $L_c$ (nm) | 24.02 |
| Hg porosimetry analysis |  |
| Specific surface area ($m^2 \cdot g^{-1}$) | 52.495 |
| Total pore volume ($cm^3 \cdot g^{-1}$) | 0.4866 |
| Porosity (%) | 36 |
| Average pore diameter (nm) | 37.1 |
| Bulk density ($g \cdot cm^{-3}$) | 0.75521 |
| Apparent density ($g \cdot cm^{-3}$) | 1.1879 |
| Energy density ($GJ \cdot m^{-3}$) | 27.97 |

The electrochemical reactions taking place in the DCFC anode occur predominantly on the carbon surface, which is generally represented by the pore walls. A higher porosity of carbon materials implies an extended specific surface area and, subsequently, more available reaction sites for electrochemical reactions and better DCFC efficiency. This provides major advantages for the biochar over other fuels with a smooth surface area, such as graphite and raw biomass. In this context, the $N_2$ BET adsorption technique was performed on the EOP biochar; unfortunately, the obtained result does not permit the prediction of its micro-structural properties. This could be due to the mesoporous and macroporous structure of the EOP biochar. Micropores will be filled in a single step over a narrow range of relative pressure before the formation of a monolayer coverage on the biochar surface. This will disturb the adsorption of $N_2$ and, subsequently, does not give useful data. This EOP biochar porous structure may be related to the operating conditions of the carbonization experiment being carried out at 400 °C for four hours. It is evident that, as carbonization temperature increases, pore blocking substances are driven off or are thermally cracked, thus increasing the externally accessible surface area [70]. However, the extended carbonization holding time (4 h) used in this study can have the opposite effect since the reactions continue at the pore surface area, causing a decrease in micropores and a shift toward meso- and macropores [70].

The scanning electron microscopy (SEM) images of the EOP biochar sample are shown in Figure 5a,b. The SEM images of the biochar produced from EOP pyrolysis at 400 °C show a hardly visible porosity (Figure 5a). The presence of crystalline phases with cubic, tubular, and elongated shapes on the particle's surfaces show that the particles are rough and grainy. The pore sizes are not uniform and are in the range of tens of nanometers to microns (Figure 5b).

**Figure 5.** Scanning electron micrographs (SEM) of the EOP biochar: (**a**) at 100 μm resolution and (**b**) at 5 μm resolution.

The mercury intrusion porosimetry was used in this study as a complementary technique to obtain a better textural characterization of the EOP biochar. Figure 6 displays the log differential mercury intrusion volume as a function of the pore diameter of the EOP biochar. Indeed, the EOP biochar presents a porosity in the mesopore–macropore range; more specifically, it is an inter-particle porosity due to the high pore size diameter (6 nm up to 3000 nm) (Figure 5), at which the mercury intrusion occurs [71]. This finding is aligned with the SEM micrographs presented above. The pore structure and the pore size distributions are summarized in Table 3. The recorded relatively low surface area observed for the EOP biochar (52.495 $m^2 \cdot g^{-1}$) is probably due to the inorganic materials, mainly silica particles, that partially fill or block the micropores.

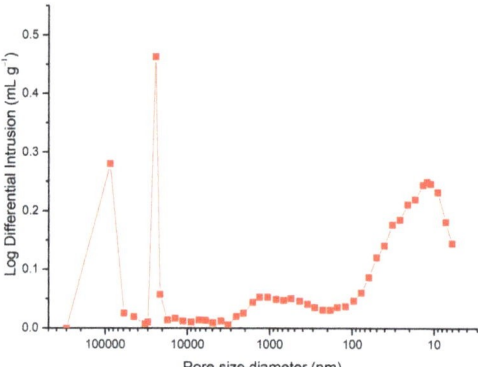

**Figure 6.** Log differential intrusion as a function of the pore diameter of the EOP biochar.

Temperature programmed oxidation (TPO) is an efficient tool to evaluate the relative activity of carbon oxidation. The weight loss curve of the EOP biochar is shown in Figure 7.

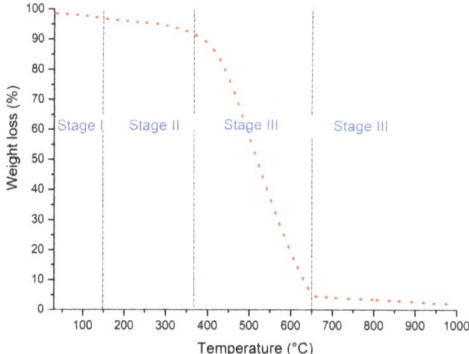

**Figure 7.** Temperature programmed oxidation (TPO) of the EOP biochar.

Based on the TPO results of EOP biochar, four stages are depicted, indicating the presence of four forms of carbons with different resistance to oxidation. The initial small weight loss below 150 °C is caused by the desorption of physiosorbed water within the EOP biochar and the oxidation of volatile organic C (Stage I). A small weight loss is observed below 380 °C due to the rapid release of combustion gases from the superficial oxygen functional groups (Stage II). This could be related to the oxidation of labile organic carbons, such as aliphatic carbon. In addition, a significant weight loss in the EOP biochar starts from 380 °C and ends at around 650 °C (Stage III). During this stage, the oxidation of recalcitrant organic carbons (predominantly formed by lignin and recalcitrant carbon, such as aromatic carbon) and refractory organic carbons (poly-condensed forms of aromatic carbon) occurs between 380 and 475 °C and between 475 and 650 °C, respectively. From 650 °C to 1000 °C, inorganic carbons (carbonates) are oxidized (Stage IV).

Compared to the TPO curves of graphite, the onset of stage III of the EOP biochar is much lower than that of graphite, which starts at 650 °C [72]. Moreover, the offset of this stage for EOP biochar ends earlier compared to graphite (850 °C). The enhanced thermal stability of the EOP biochar in air can be assigned to its lower graphitic degree and crystallinity, which confirms the findings observed in its XRD analysis.

FTIR spectroscopy was performed in order to analyze the surface functional groups of the EOP biochar, aiming to assess its stability. The EOP biochar FTIR spectrum is shown in Figure 8.

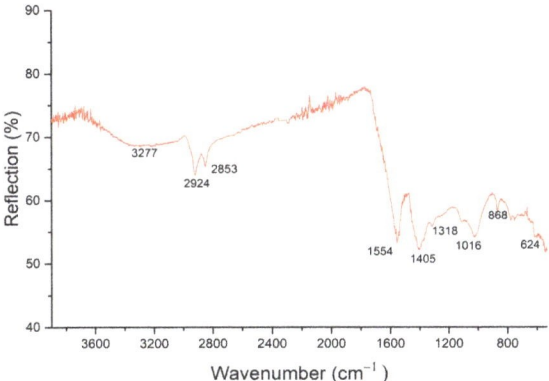

**Figure 8.** FTIR spectrum of the EOP biochar sample.

The broad band at 3400–3200 cm$^{-1}$ indicates the presence of O-H stretching vibration. The bands at the region between 3000 and 2800 cm$^{-1}$ are related to the presence of aliphatic C-H stretching vibrations. The bands at 1600–1500 cm$^{-1}$ denote the presence of aromatic C=C ring stretching. The bands at 1405 cm$^{-1}$ and 1318 cm$^{-1}$ are assigned to the O-H bending vibrations in acid, alcohols, and phenol groups. The band at 1016 cm$^{-1}$ indicates the C-O stretching vibrations in carbonyl compounds, such as alcohols, phenols, ester, ether, and acid. The bands at 900–600 cm$^{-1}$ are related to the C-H bending vibrations in aromatic hydrocarbons. This analysis shows that the derived EOP biochar structure is mainly composed of aromatics, which explain the lower value of the H/C ratio reported in Table 1.

### 3.3. Feasibility Testing of the EOP Biochar as Fuel in DCFC

The direct carbon fuel cell voltage versus current density curve and the power density versus current density curve of the EOP biochar are shown in Figure 9, in comparison to the performances recorded for the DCFCs fed by biochars from almond shell (AS) [19] and olive wood (OW) [29] that were prepared under the same carbonization process.

The DCFC pellets used for testing the electrochemical performance of the EOP biochar sample are also the same as those used for the AS and OW biochar tests in terms of electrolyte and cathode materials, cell geometry, gas flow rates, and operating temperature (700 °C) [19,29]. On the basis of the experimental observation after the cell tests, in the case of the EOP biochar, the fuel utilization must be reduced in comparison to the case of the AS and OW biochars tested in previous works. This could be related to the high content of ash. This result is in agreement with the literature [73,74]. In the case of corn straw carbon, Li et al. [73] noted a better fuel utilization rate and a lower weight of remains after discharging in the DCFC anode, which contains only 1.6 wt% of ash.

The obtained results prove the feasibility of DCFC operation by using the EOP biochar as a fuel, showing an open circuit voltage of 0.8 V, a peak power density $P_{max}$ of about 10 mW·cm$^{-2}$, and a limiting current density of 142 mA·cm$^{-2}$ at 700 °C. The main useful criterion of the evaluation of DCFC electrochemical performance and quality is the achieved maximum power density. The recorded $P_{max}$ of about 10 mW·cm$^{-2}$ in this study can be considered as promising if compared to the reported power densities of several biochars from pistachio shells (3.7 mW·cm$^{-2}$), pecan shells (3.2 mW·cm$^{-2}$), and sawdust (1.4 mW·cm$^{-2}$), which were all tested in DCFCs at 700 °C [31]. Nevertheless, this performance appears so modest compared to those obtained by the AS and OW biochar-fed DCFCs reported in Figure 9, which exhibited a performance 10 times higher than that of the EOP biochar.

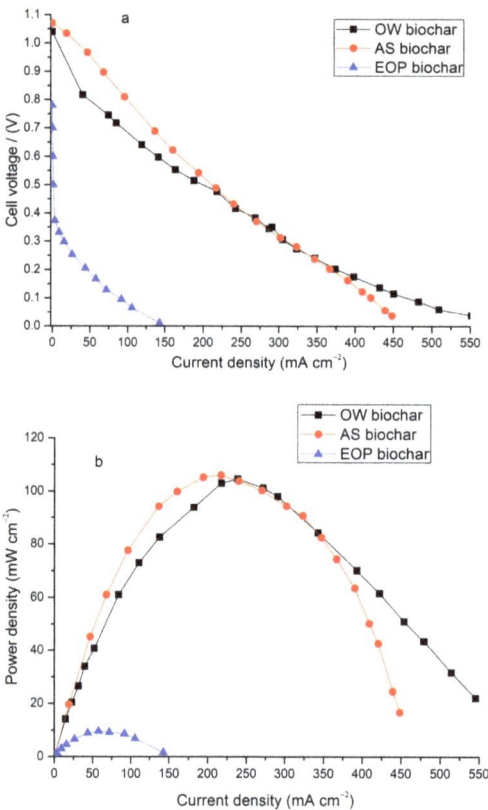

**Figure 9.** (**a**) I–V and (**b**) I–P curves of the DCFC powered by the EOP biochar at 700 °C, compared to those of the DCFCs powered by AS biochar [19] and OW biochar [29].

It is worth highlighting that four peculiar phenomena are observed in the recorded performance curve, including the low OCV value, the high activation loss at low current densities, the limited peak power density ($P_{max}$), and the reduced limiting current density ($i_{lim}$). We have to address the physico-chemical properties of the EOP-derived biochar sample that are certainly the cause of these phenomena, as previous characterization have demonstrated some peculiar properties when compared to other biochars prepared under the same carbonization conditions, such as AS and OW biochars [19,29]. This difference is certainly related to the raw biomass's material properties and the kinetics of the complex chemical degradation reactions occurring during the carbonization process.

Several researchers proved that the overall DCFC efficiency is ascribed mainly to the gas–active electrochemical zone interactions, rather than to the extremely pure carbon–active electrochemical zone contact. The reverse Boudouard reaction (Equation (5)), which represents the main non-electrochemical reaction possible to occur at high DCFC temperatures (>700 °C), has a determining role in the DCFC performance as its gaseous product (CO) can readily diffuse within the active electrochemical reaction site at a considerably faster rate than solid carbon, contributing highly to the enhancement in the power output. In addition to CO gas, the in situ generation of hydrogen ($H_2$) is another key point when the DCFC is fed by hydrogenated carbonaceous materials and is also a cause of the high values of DCFC power densities.

In this context, one of the carbonaceous material's properties that can affect the $P_{max}$ recorded by DCFCs is the carbon content. A biochar containing low carbon generates low amounts of obstructive matters (CO and $CO_2$), which potentially affects the pathways of chemical and electrochemical reactions occurring in the cell anode. The carbon content in the EOP biochar sample (65.7%) is lower than that of the AS biochar (71.8%) [19] and that of the OW biochar (84.07%) [29], which could be one of the causes of the limited performance recorded by the DCFC fed by the EOP biochar. Indeed, the EOP low carbon content may hinder and reduce the amount of $CO_2$ and CO gases produced in situ at the anode compartment, subsequently limiting the extent of the CO electrochemical reactions within the anode of the DCFC (Equations (9) and (10)).

Additionally, another biochar property, which may explain the limited power density recorded in this case study, is related to the proximate composition, which has a direct effect on the evolved gases emitting off from the carbonaceous skeleton via thermal cracking reactions. Nevertheless, the literature shows that there is no consensus on the influence of biochar volatile matter content on DCFC performance. Chien et al. [75] reported that a high content of volatile matter causes a limitation in DCFC performance, while a medium volatile matter content is more appropriate for the use of a biochar as fuel in the DCFC. In contrast, Kaklidis et al. [30] affirmed that the volatile matter content greatly improves the electrochemical output of DCFCs.

Fuel ratio is a ratio of the amount of fixed carbon to volatile matter and is a distinctive feature of solid fuels, which can be used to categorize them as a coal rank according to the ASTM 388 [33]. Compared to the OW and AS biochars [19,29], which have a respective fuel ratio of about 0.563 and 0.245, the EOP biochar has a higher fuel ratio (0.897) as its fixed carbon is relatively high (about 36.45%), whereas the volatile matter is limited (40.62%). The improved fuel ratios of biochars are known to be significant, indicating greater combustion efficiencies and reduced pollutant emissions during the burning process of biochars, when compared to the burning of their raw forms. Contrarily, it appears that in the case of DCFCs, lower fuel ratio in the biochars is a preferred property for better cell output.

Additionally, the original lignocellulosic composition of the biomass can provide insight into the mass yield of the produced biochar through pyrolysis and, subsequently, its performance as a fuel in a DCFC system.

According to the DCFC performance recorded for the EOP biochar, the biomass composition is dominated by lignin, rather than by holocellulosic content, which is not preferred despite its potential in increasing the yields of the biochar. Correspondingly, the high ash content in the biomass favors charring reactions rather than devolatilization reactions [76], leading to an enhancement in the yield of the biochar.

Seeing that the objective is not focused on the biochar quantity produced from a biomass but rather its quality and potential to be electrochemically active, leading to a promising cell output, we notice that a biomass with a high lignin and ash content is not preferred in the production of a biochar fuel for DCFC systems.

Moreover, the crystallographic structure of carbonaceous materials is a key property over others that directly affects the electrical conductivity and the reactivity of biochars. A high degree of order in the structure provides enhanced electrical conductivity. We should keep in mind that when carbon is applied as fuel in a DCFC, the reactivity, which is known to be higher in non-graphitic structures, is more important than high electrical conductivity values [77]. It is often stated in the literature that the electrical conductivity of solid carbonaceous materials is not correlated with the DCFC power output, regardless of the diversity of anode configurations. Logically, higher electrical conductivity values are generally recorded for graphitic or crystalline structures of carbon materials, which should induce a reduced global DCFC ohmic resistance and, consequently, a better DCFC performance. Unfortunately, that has never been observed when these high-ordered carbon materials were employed as fuels in DCFC systems. The need for crystalline carbon materials was revealed when using carbon materials simultaneously as the anode, the current collector, and the fuel in a few DCFC configurations. In contrast, when using an anode current collector (silver for example) to collect electrons, the dependence of the cell performance on the electrical conductivity of carbon materials was more curtailed.

Meanwhile, the crystalline structure of solid carbon materials has often been measured using XRD analysis. Chien et al. [41] reported that a DCFC's electric power generation and long-term stability have a strong correlation with the crystalline structure of solid carbons. A high power density and good stability are obtained when using less crystalline carbon, which is characterized by a broad (002) diffraction peak et $2\theta = 20$–$30°$ on the XRD pattern.

In this study, the obtained XRD pattern of the EOP biochar demonstrates a high number of peaks, which is different to the AS [19] and OW [29] biochar patterns that exhibit only two peaks. As previously mentioned, the EOP biochar has a high ash content, which explains the diversity of peaks recorded as belonging to silica, dolomite, sylvite, etc. This represents one of the features of a low reactive carbon material and explains the low recorded peak power density value. Based on the limited performance recorded by the DCFC fed by the EOP biochar (Figure 9), it is shown that the biochar's disordered structure is not sufficient to judge its potential as fuel in a DCFC system, but it is necessary to assess more of its content in terms of mineral crystalline materials.

More specifically, the presence of silica is a cause of the low EOP biochar reactivity and conductivity to electrons because the silica non-crystalline particles act as insulators to electron transfer and as inhibitors to electrochemical oxidation reactions.

As previously mentioned, the reactivity of carbon materials appears to be a key property affecting the maximum power value ($P_{max}$) generated by the DCFC. This is evidenced through the results obtained by TPO analysis. The reactivity of the EOP biochar is also lower than that of the AS and OW biochars and can be considered as another reason for the limited DCFC performance shown in Figure 9; stage III of the EOP biochar's TPO profile (Figure 6) starts at almost the same temperature as the AS and OW biochars but ends later at about 650 °C in comparison to 580 °C and 590 °C in the case of the AS and OW biochars, respectively.

Eom et al. [78] affirmed that the ratio of total oxygen to carbon, i.e., surface functional groups containing oxygen, has the dominant effect on the electrochemical reaction relative to the surface area in a DCFC. However, in our case, the EOP exhibits a higher O/C ratio, which means higher surface oxygen groups, compared to the AS and OW biochars [19,29], but the DCFC performance is comparatively limited. It is also known that when the amount of oxygen-containing surface groups increases remarkably, it favors a decrease in the electrical conductivity due to the preponderance of insulating effects caused by the functional groups on the surface [79].

Regarding the hybrid character of the DCFC anode, which allows the simultaneous electrochemical conversion of both solid carbon and reformed gaseous fuels (CO and $H_2$), an analysis of the nature of the surface oxygenated functional groups, apart from their extent, enables an estimation of their production potential of the gaseous species mentioned above and, subsequently, allows an assessment of their contribution to the complex reaction scheme of anode chemistry and electrochemistry. The FTIR spectrum of the EOP biochar (Figure 8) indicates the absence of the peak of C=O at about 1700–1710 cm$^{-1}$, which reflects the absence of carbonyl, carboxyl, lactone, and anhydride groups within the surface chemical structure of the EOP biochar. These oxygen functional groups are known to contribute to the generation of CO and $CO_2$ gases within the anode and, thus, cause the occurrence of several chemical reactions, such as the water–gas shift reaction, the Boudouard reaction, the water shift reaction, and the methanation reaction. Elleuch et al. [29] proposed a chemical mechanism of OW biochar devolatilization within a DCFC anode and showed that $CO_2$ evolves from the decomposition of carboxylic acid functionality at low temperatures and/or lactones at high temperatures, while CO arises from carbonyls at high temperatures.

On the other hand, the EOP biochar's FTIR spectrum (Figure 8) shows the presence of C-O single-bond groups ascribed to the high content of phenols and ethers. These groups are more difficult to decompose but are able to produce CO and $CO_2$ at very high temperatures in the presence of catalysts. In conclusion, it seems that the existence of C=O bonds is preferred over C-O bonds for the easiest way to generate gases within the anode of a DCFC and for the further electrochemical contribution of these gases to the DCFC electrochemistry; these bonds are absent in the case of the EOP biochar. Moreover, the high O/C content indicates the extended availability of surface oxygen groups, although this information remains not informative. We need to identify the chemical functional groups that are known to be easily decomposed, such as carbonyl groups.

The DCFC fed by the EOP biochar delivers a limited current density, which is also correlated with its distinct physico-chemical properties. HHV is a fuel property that indicates the chemical energy density that will be converted into electricity through DCFC systems. The HHV of the EOP biochar is higher than that of the EOP raw biomass due to the conversion of oxygenated compounds into carbonaceous hydrocarbons, which reduces the oxygen content in the biochar and increases the carbon content. As shown in Figure 9, the limiting current density delivered by the DCFC fueled by the EOP biochar (142 mA·cm$^{-2}$) is so low compared to that of the AS biochar (450 mA·cm$^{-2}$) and the OW biochar (550 mA·cm$^{-2}$). The HHV also follows the same trend: OW biochar (27.31 MJ·kg$^{-1}$) [29] > AS biochar (24.7 MJ·kg$^{-1}$) [19] > EOP biochar (22.21 MJ·kg$^{-1}$). It is, thus, evident that the higher the biochar energy density is, the higher the current density delivered by the cell is; however, there is a high disproportion between the ordered HHV values and the decrease in the limiting current density.

In order to investigate this behavior between the HHV and the limiting current of the cell, we selected another biochar (corn cob, noted as CC) tested in a similar DCFC system by Yu et al. [49]. The results illustrated in Figure 10a show that the limiting current density recorded by the DCFC follows a polynomial trend as a function of the biochar's HHVs, according to the following equation:

$i_{lim}$ = −14.955 HHV$^2$ + 821.23 HHV − 10,723 with R$^2$ = 0.9962. This correlation could predict the DCFC's output when applying different types of carbon materials as fuel in similar DCFC operating conditions (700 °C) and materials.

As the biochar's HHVs range from 22 to 32.5 MJ·kg$^{-1}$, we used the previously mentioned polynomial law to assess the dependency between the biochar's property HHV and its impact on the cell's current output. The obtained curve (Figure 10b) shows that the biochar has a HHV ranging between 24 and 30 MJ·kg$^{-1}$, resulting in a promising DCFC current density (higher than 400 mA·cm$^{-2}$).

The electrochemical reactions in a DCFC system occur predominantly on the carbon surface represented by the pores walls. Consequently, a higher porosity results in a larger surface area. It is known that a high interfacial surface enlarges the availability of sites for electrochemical reactions and, consequently, leads to an increase in the electron flux generated via the electrochemical reactions of the DCFC.

Figure 6 shows the distribution of the pore diameter within the EOP biochar, which is similar to that of the AS and OW biochars but different from derived biochars from agricultural wastes, which are often considered to be entirely microporous (D < 2 nm). The presence of the meso–macroporosity in the EOP biochar helps ameliorate the mass transport and the kinetic processes, which results in the promising current density output recorded when testing the EOP biochar (142 mA·cm$^{-2}$).

The surface area of a wooden biochar is about 131.4 m$^2$·g$^{-1}$ [17], which is limited but higher than that of the analyzed EOP biochar (52.495 m$^2$·g$^{-1}$). These surface area values are in good agreement with those reported in the literature, especially with the fact that these samples have resulted from a carbonization process with no further activation step [71]. We should also notice that the AS and OW biochars exhibit very limited specific surface areas in the range between 30 and 40 m$^2$ g$^{-1}$, which is in a comparable range with that of the EOP biochar sample.

There is a strong correlation between the pore volume and surface area measurements based on N$_2$ adsorption, but the use of pore volume may be preferable due to the low surface area values in the case of biochar obtained from carbonization [41].

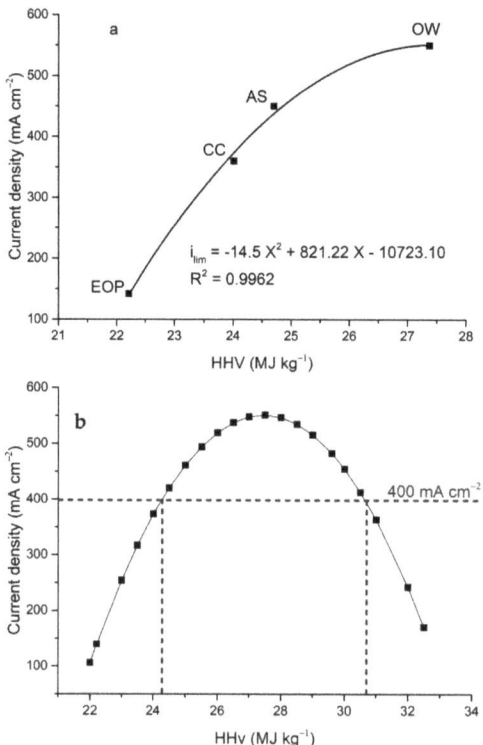

**Figure 10.** Evolution of (**a**) experimentally recorded current densities and (**b**) modeled polynomial current densities delivered by the DCFC system as a function of the biochar's HHVs.

The total pore volume of the EOP biochar is three and two times lower than the OW and AS biochars, respectively. Similarly, its porosity is very limited (about 36%). This may be caused by the high content of silica within the EOP biochar sample, which hinders its porousness and leads to the limited current density ($i_{lim}$) delivered by the DCFC (Figure 9). The $i_{lim}$ also depends on the diffusion ability of species, which appears to be restricted in this case. Porousness could, thus, be enhanced by removing silica from the biochar, which can be transformed into activated carbon when exposed to higher temperature.

The open circuit voltage (OCV) of the DCFC fed by the EOP biochar is lower than the theoretical OCV value (ca. 1.02 V) of the DCFC. This low OCV is not related to the low densification of the composite electrolyte, which is proven through an SEM investigation of the electrolyte after the test, as shown in Figure 11. In Figure 11a, the porous structure of the DCFC pellet at room temperature is clearly observed. The composite electrolyte and the cathode layers indicate different morphologies. A large particle size distribution ranging from 90 to 500 nm and from 45 to 312 nm for the $Li_xNi_{1-x}O$ cathode particles and the composite electrolyte particles, respectively, are obtained. This is not the case in the second micrograph of the DCFC pellet obtained after the electrochemical test (Figure 11b). After the DCFC test, it becomes difficult to distinguish between the pellet layers (cathode/electrolyte). Additionally, a distinction of the SDC particles from the carbonate phase is not possible because the SDC particles are coated completely by the solidified carbonate, which has once melted, proving the fact that the electrolyte is well densified and prevents the cathode gas from leaking through during the DCFC electrochemical test. The composite electrolyte provides the pathways for the transfer of both oxide and carbonate ions. Thus, the nature of the fuel may be the cause of this OCV tendency. This behavior is different from that recorded in our previous studies using the AS [19] and OW [29] biochars. This deviation can be ascribed to the distinct and complex schemes of chemical and electrochemical reactions occurring within the anodic side, yielding a lower concentration of reactive gaseous species (CO and $H_2$) (Figure 12).

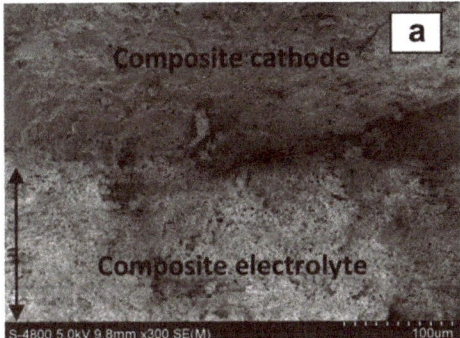
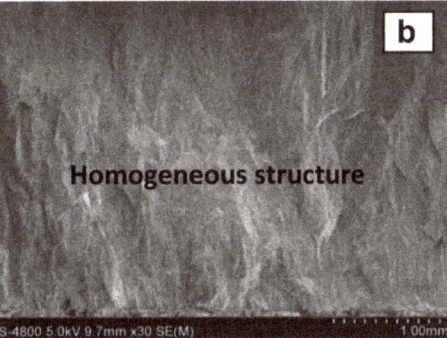

**Figure 11.** Scanning electron micrograph of the electrolyte structure: (**a**) before cell testing at room temperature and (**b**) after cell testing at T = 700 °C.

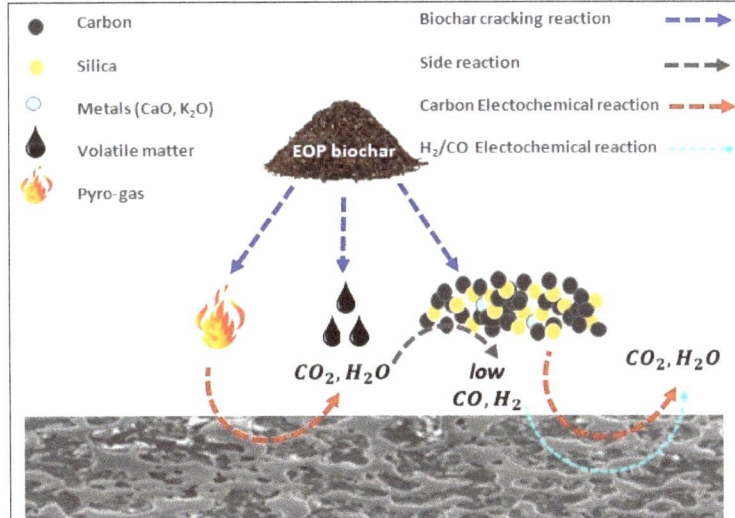

**Figure 12.** Presentation of the anodic reaction mechanisms of the EOPbiochar as a fuel in the DCFC.

The limited OCV value is, thus, another proof of the previously outlined finding related directly to the reduced formation of gases as fuels due to the absence of carbonyl groups at the surface functionalities of the EOP biochar. This is aligned with the results reported by Kaklidis et al. [30], who clearly correlated the CO formation rate obtained under open circuit conditions with the achieved DCFC electrochemical performance and proved that the overall chemical and electrochemical processes are driven by the CO shuttle mechanism.

Furthermore, it appears that other physico-chemical properties of the biochar, such as the mineral composition, strongly affect the gas phase/surface chemical reactions and the charge transfer phenomena, modifying the anode gas composition and, subsequently, the recorded OCV.

In the case of DCFC application, the research work by Li and co-workers [43], who studied the effect of several metallic oxides common to coal ash on the electrochemical performance of DCFCs through the addition of a fixed amount (8%) of the oxides of calcium, magnesium, iron, aluminum, and silica impregnated in activated carbon (AC), confirmed that MgO, CaO, and $Fe_2O_3$ act as catalysts for the electrochemical process of DCFCs [43]. Elleuch et al. [29] analyzed an OW biochar's performance as a fuel in a DCFC system and affirmed that CaO and $K_2O$ act as catalysts for the anodic electrochemical mechanism of the cell. The same catalytic effect of the latter alkali and alkaline metals ($K_2O$ and CaO) was also revealed in the case of an AS biochar-fed DCFC, which is considered one of the main positive characters of the biochar, leading to the promising performance. We should also notice that these minerals are able to catalyze several devolatilization reactions occurring within the biochar skeleton, such as the Boudouard reaction and the water–gas shift reaction.

In this study, the EOP biochar's content of $K_2O$ (13.72 wt%) and CaO (16.31 wt%) minerals in the inorganic fraction of the ash is far different to both the OW and AS biochars, which were previously affirmed as promising fuels in DCFCs [19,29]. This difference is primarily related to the biomass essence.

Additionally, the catalytic contribution of these metals to the DCFC's electrochemical and the biochar's devolatilization reactions seems to be curtailed by a high content of silica in the EOP biochar. This is in agreement with other research works on gasification catalysts, which concluded that the catalytic effect of metals, such as (K), was reduced by the reaction with silica to form silicate during gasification, rather than to the accelerated devolatilization reactions [64].

Another peculiarity observed in the I–V curve of the DCFC fed by the EOP biochar (Figure 9) is related to the pronounced activation polarization. This voltage loss results from the difficulties encountered by reactive species (solid carbon, CO, and $H_2$) when carrying out electrochemical reactions at lower current densities. The limited kinetics of reactions is also related to the physico-chemical properties of the biochar. Biochar porousness will curtail the reaction sites, and a high silica content will hinder the diffusion of species on the one hand and contribute to the insulation of some of the available reaction sites on the other hand. Altogether, these explain the high activation polarization taking place during the operation of the DCFC fed by the EOP biochar.

## 4. Conclusions

In this work, the electrochemical performance of a biochar produced from EOP biomass carbonization at 400 °C was examined as a fuel in a DCFC system supported by a 0.6 mm-thick $Ce_{0.8}Sm_{0.2}O_{1.9}$ (SDC)-carbonate composite electrolyte layer. The cell exhibits a maximum power density of about 10 mW·$cm^{-2}$ and a limiting current density of about 142 mA·$cm^{-2}$ at 700 °C, which is limited if compared to the AS and OW biochar-fed DCFCs prepared under similar conditions. The OCV is about 0.87 V lower compared to the DCFC's theoretical voltage. This is ascribed to the various formed gases emitted from the porous biochar structure, which plays a key role in the DCFC's performance apart from the solid carbon skeleton. Even though the electro-oxidation of solid carbon and volatile gases has a significant effect on the performance of the DCFC, the contribution of each one has not been fully assessed. The physico-chemical and structural properties of the EOP biochar were analyzed using elemental and proximate analyses, mercury porosimetry, scanning electron microscopy, X-ray diffraction, FTIR analysis, and X-ray fluorescence analysis. The thermogravimetric analysis under a normal air atmosphere enables the conclusions about the thermal stability and reactivity of the EOP biochar. On the basis of this extensive characterization study, a direct correlation between the EOP biochar's physico-chemical characteristics and the power output is revealed.

The carbon, oxygen, ash (impurities), and volatile matter contents; the porosity and surface area; the presence of carbonyl/carboxylic groups; and the disorder of the carbon structure exert a pronounced impact on the electrochemical performance of the EOP biochar-fed DCFC.

An enhanced electrochemical reactivity of the EOP biochar in the DCFC system could be achieved by first removing silica from the biochar using a demineralization pre-treatment, followed by activation.

The present research investigated the positive and negative properties affecting the performance of a biochar as fuel in a DCFC. The major conclusions show that a promising biochar fuel for DCFCs is one that comes from a biomass feedstock with a high HHV, a high holocellulose/lignin ratio, and, especially, a low ash content, without looking at its mineral composition.

**Author Contributions:** Experimental work and writing—original draft preparation: N.G. and A.E.; review and supervision: A.E., K.H. and Y.L. All authors have read and agreed to the published version of the manuscript.

**Funding:** This research was funded by the Ministry of Higher Education and Scientific Research of Tunisia in the framework of the ERANETMED2-72-251, Medwaste project.

**Data Availability Statement:** Additional details of the experiments and data may be asked via email to the corresponding authors.

**Acknowledgments:** This research is supported by the Tunisian Ministry of Higher Education and Scientific Research in the framework of the ERANETMED2-72-251, Medwaste project. The financial support of the NSF of China under contract number21120102039, the support of the Tianjin Municipal Science and Technology commission under contract number 13JCZDJC26600, the support of the Program of Introducing Talents to the University Disciplines under file number B06006, the support of the Program for Chengjiang Scholars and Innovative Research Teams in Universities under file number IRT 0641, and the support of the Ministry of Education of China under contract number 20130032120023 are also gratefully acknowledged.

**Conflicts of Interest:** The authors declare no conflict of interest.

## References

1. Saviozzi, A.; Levi-Minzi, R.; Cardelli, R.; Biasci, A.; Riffaldi, R. Suitability of moist olive pomace as soil amendment. *Water Air Soil Pollut.* **2001**, *128*, 13–22. [CrossRef]
2. Alburquerque, J.A.; Gonzálvez, J.; García, D.; Cegarra, J. Agrochemical characterization of "alperujo", a solid by-product of the two-phase centrifugation method for olive oil extraction. *Bioresour. Technol.* **2004**, *91*, 195–200. [CrossRef] [PubMed]
3. Gómez-Muñoz, B.; Hatch, D.J.; Bol, R.; García-Ruiz, R. The Compost of Olive Mill Pomace: From a Waste to a Resource—Environmental Benefits of Its Application in Olive Oil Groves. In *Sustainable Development—Authoritative and Leading Edge Content for Environmental Management*; Curkovic, S., Ed.; In Tech: Rijeka, Croatia, 2012; pp. 449–484, ISBN 9789535106821.
4. Zabaniotou, A.A.; Kalogiannis, G.; Kappas, E.; Karabelas, A.J. Olive residues (cuttings and kernels) rapid pyrolysis product yields and kinetics. *Biomass Bioenergy* **2000**, *18*, 411–420. [CrossRef]
5. Jauhiainen, J.; Conesa, J.A.; Font, R.; Martín-Gullón, I. Kinetics of the pyrolysis and combustion of olive oil solid waste. *J. Anal. Appl. Pyrolysis* **2004**, *72*, 9–15. [CrossRef]
6. Naik, S.N.; Goud, V.V.; Rout, P.K.; Dalai, A.K. Production of first and second generation biofuels: A comprehensive review. *Renew. Sustain. Energy Rev.* **2010**, *14*, 578–597. [CrossRef]
7. Volpe, M.; D'Anna, C.; Messineo, S.; Volpe, R.; Messineo, A. Sustainable production of bio-combustibles from pyrolysis of agro-industrial wastes. *Sustainability* **2014**, *6*, 7866–7882. [CrossRef]
8. Miranda, T.; Nogales, S.; Román, S.; Montero, I.; Arranz, J.I.; Sepúlveda, F.J. Control of several emissions during olive pomace thermal degradation. *Int. J. Mol. Sci.* **2014**, *15*, 18349–18361. [CrossRef] [PubMed]
9. Miranda, I.; Simões, R.; Medeiros, B.; Nampoothiri, K.M.; Sukumaran, R.K.; Rajan, D.; Pereira, H.; Ferreira-Dias, S. Valorization of lignocellulosic residues from the olive oil industry by production of lignin, glucose and functional sugars. *Bioresour. Technol.* **2019**, *292*, 121936. [CrossRef] [PubMed]
10. Cuevas, M.; García, J.F.; Hodaifa, G.; Sánchez, S. Oligosaccharides and sugars production from olive stones by autohydrolysis and enzymatic hydrolysis. *Ind. Crops Prod.* **2015**, *70*, 100–106. [CrossRef]
11. Pütün, A.E.; Uzun, B.B.; Apaydin, E.; Pütün, E. Bio-oil from olive oil industry wastes: Pyrolysis of olive residue under different conditions. *Fuel Process. Technol.* **2005**, *87*, 25–32. [CrossRef]
12. Chouchene, A.; Jeguirim, M.; Khiari, B.; Zagrouba, F.; Trouvé, G. Thermal degradation of olive solid waste: Influence of particle size and oxygen concentration. *Resour. Conserv. Recycl.* **2010**, *54*, 271–277. [CrossRef]

13. Özveren, U.; Özdoğan, Z.S. Investigation of the slow pyrolysis kinetics of olive oil pomace using thermo-gravimetric analysis coupled with mass spectrometry. *Biomass Bioenergy* **2013**, *58*, 168–179. [CrossRef]
14. Parascanu, M.M.; Sánchez, P.; Soreanu, G.; Valverde, J.L.; Sanchez-Silva, L. Environmental assessment of olive pomace valorization through two different thermochemical processes for energy production. *J. Clean. Prod.* **2018**, *186*, 771–781. [CrossRef]
15. Dinc, G.; Yel, E. Self-catalyzing pyrolysis of olive pomace. *J. Anal. Appl. Pyrolysis* **2018**, *134*, 641–646. [CrossRef]
16. Cao, D.; Sun, Y.; Wang, G. Direct carbon fuel cell: Fundamentals and recent developments. *J. Power Source* **2007**, *167*, 250–257. [CrossRef]
17. Ahn, S.Y.; Eom, S.Y.; Rhie, Y.H.; Sung, Y.M.; Moon, C.E.; Choi, G.M.; Kim, D.J. Utilization of wood biomass char in a directcarbon fuel cell (DCFC) system. *Appl. Energy* **2013**, *105*, 207–216. [CrossRef]
18. Cherepy, N.J.; Krueger, R.; Fiet, K.J.; Jankowski, A.F.; Cooper, J.F. Direct Conversion of Carbon Fuels in a Molten CarbonateFuel Cell. *J. Electrochem. Soc.* **2005**, *152*, A80–A87. [CrossRef]
19. Elleuch, A.; Boussetta, A.; Yu, J.; Halouani, K.; Li, Y. Experimental investigation of direct carbon fuel cell fueled by almond shell biochar: Part I. Physico-chemical characterization of the biochar fuel and cell performance examination. *Int. J. Hydrogen Energy* **2013**, *38*, 16590–16604. [CrossRef]
20. Heneka, M.J.; Ivers-Tiffée, E. Degradation of SOFC single cells under severe current cycles. *Proc. Electrochem. Soc.* **2005**, *2005*, 534–543. [CrossRef]
21. Yamaji, K.; Ishiyama, T.; Bagarinao, K.D.; Kishimoto, H.; Horita, T.; Yokokawa, H. Evaluation of Impurity Levels in Cathods of Seven Different SOFCS tacks and Modules before and after Long-Term Operation. *ECS Meet. Abstr.* **2017**, *MA2017-03*, 177. [CrossRef]
22. Jiang, C.; Ma, J.; Arenillas, A.; Bonaccorso, A.D.; Irvine, J.T.S. Comparative study of durability of hybrid direct carbon fuelcells with anthracite coal and bituminous coal. *Int. J. Hydrogen Energy* **2016**, *41*, 18797–18806. [CrossRef]
23. Alenazey, F.; Alyousef, Y.; Alotaibi, B.; Almutairi, G.; Minakshi, M.; Cheng, C.K.; Vo, D.V.N. Degradation Behaviors of Solid Oxide Fuel Cell Stacks in Steady-State and Cycling Conditions. *Energy Fuels* **2020**, *34*, 14864–14873. [CrossRef]
24. Xia, C.; Li, Y.; Tian, Y.; Liu, Q.; Wang, Z.; Jia, L.; Zhao, Y.; Li, Y. Intermediate temperature fuel cell with adoped ceria-carbonate composite electrolyte. *J. Power Source* **2010**, *195*, 3149–3154. [CrossRef]
25. Zhao, Y.; Xia, C.; Wang, Y.; Xu, Z.; Li, Y. Quantifying multi-ionic conduction through doped ceria-carbonate composite electrolyte by a current-interruption technique and product analysis. *Int. J. Hydrogen Energy* **2012**, *37*, 8556–8561. [CrossRef]
26. Jia, L.; Tian, Y.; Liu, Q.; Xia, C.; Yu, J.; Wang, Z.; Zhao, Y.; Li, Y. A direct carbon fuel cell with (molten carbonate)/(doped ceria) composite electrolyte. *J. Power Source* **2010**, *195*, 5581–5586. [CrossRef]
27. Cui, C.; Li, S.; Gong, J.; Wei, K.; Hou, X.; Jiang, C.; Yao, Y.; Ma, J. Review of molten carbonate-based direct carbon fuel cells. *Mater. Renew. Sustain. Energy* **2021**, *10*, 1–24. [CrossRef]
28. Mejdoub, F.; Elleuch, A.; Halouani, K. Assessment of the intricate nickel-based anodic reactions mechanism within a methanol fed solid oxide fuel cell based on a co-ionic conducting composite electrolyte. *J. Power Source* **2019**, *414*, 115–128. [CrossRef]
29. Elleuch, A.; Halouani, K.; Li, Y. Investigation of chemical and electrochemical reactions mechanisms in a direct carbon fuel cell using olive wood char coal as sustainable fuel. *J. Power Source* **2015**, *281*, 350–361. [CrossRef]
30. Kaklidis, N.; Kyriakou, V.; Garagounis, I.; Arenillas, A.; Menéndez, J.A.; Marnellos, G.E.; Konsolakis, M. Effect of carbon type on the performance of a direct or hybrid carbon solid oxide fuel cell. *RSC Adv.* **2014**, *4*, 18792–18800. [CrossRef]
31. Konsolakis, M.; Kaklidis, N.; Marnellos, G.E.; Zaharaki, D.; Komnitsas, K. Assessment of biochar as feedstock in a direct carbon solid oxide fuel cell. *RSC Adv.* **2015**, *5*, 73399–73409. [CrossRef]
32. Chen, C.C.; Selman, J.R. Anode modeling of a molten-carbonate based direct carbon fuel cell. *J. Power Source* **2017**, *353*, 312–322. [CrossRef]
33. Jafri, N.; Wong, W.Y.; Doshi, V.; Yoon, L.W.; Cheah, K.H. A review on production and characterization of biochars for application in direct carbon fuel cells. *Process Saf. Environ. Prot.* **2018**, *118*, 152–166. [CrossRef]
34. Wang, J.; Fan, L.; Yao, T.; Gan, J.; Zhi, X.; Hou, N.; Gan, T.; Zhao, Y.; Li, Y. A High-Performance Direct Carbon Fuel Cell with Reed Rod Biocharas Fuel. *J. Electrochem. Soc.* **2019**, *166*, F175–F179. [CrossRef]
35. Shama, P.; Singh, D.; Minakshi, M.; Quadsia, S.; Ahuja, R. Activation-Induced Surface Modulation of Biowaste-Derived Hierarchical Porous Carbon for Supercapacitors. *ChemPlus Chem.* **2022**, *87*, e202200126.
36. Wickramaarchchi, K.; Minakshi, M.; Aravindh, S.A.; Dabare, R.; Gao, X.; Jiang, Z.T.; Wong, K.W. Repurposing N-Doped Grape Marcfor the Fabrication of Supercapacitors with Theoretical and Machine Learning Models. *Nanomaterials* **2022**, *12*, 1847. [CrossRef]
37. Gómez, I.C.; Cruz, O.F.; Silvestre-Albero, J.; Rambo, C.R.; Escandell, M.M. Role of KCl in activation mechanisms o KOH-chemically activated high surface area carbons. *J. $CO_2$ Util.* **2022**, *66*, 102258. [CrossRef]
38. Jayakumar, A.; Küngas, R.; Roy, S.; Javadekar, A.; Buttrey, D.J.; Vohs, J.M.; Gorte, R.J. A direct carbon fuel cell with a molten antimony anode. *J. Energy Environ. Sci.* **2011**, *4*, 4133–4137.
39. Cai, W.; Zhou, Q.; Xie, Y.; Liu, J. A facile method of preparing Fe-loaded activated carbon fuel for direct carbon solid oxide fuel cells. *Fuel* **2015**, *159*, 887–893. [CrossRef]
40. Hao, W.; Mi, Y. Evaluation of waste paper as a source of carbon fuel for hybrid direct carbon fuel cells. *Energy* **2016**, *107*, 122–130. [CrossRef]

41. Chien, A.C.; Chuang, S.S.C. Effect of gas flow rates and Boudouard reactions on the performance of Ni/YSZ anode supported solid oxide fuel cells with solid carbon fuels. *J. Power Source* **2011**, *196*, 4719–4723. [CrossRef]
42. Hao, W.; He, X.; Mi, Y. Achieving high performance in intermediate temperature direct carbon fuel cells with renewable carbon as a fuel source. *J. Appl. Energy* **2014**, *135*, 174–181. [CrossRef]
43. Li, X.; Zhu, Z.; De Marco, R.; Bradley, J.; Dicks, A. Evaluation of raw coals as fuels for direct carbon fuel cells. *J. Power Source* **2010**, *195*, 4051–4058. [CrossRef]
44. Kacprzak, A.; Kobyłecki, R.; Włodarczyk, R.; Bis, Z. Efficiency of non-optimized direct carbon fuel cell with molten alkaline electrolyte fueled by carbonized biomass. *J. Power Source* **2016**, *321*, 233–240. [CrossRef]
45. Wang, C.Q.; Liu, J.; Zeng, J.; Yin, J.L.; Wang, G.L.; Cao, D.X. Significant improvement of electro oxidation performance of carbon in molten carbonates by the introduction of transition metal oxides. *J. Power Source* **2013**, *233*, 244–251. [CrossRef]
46. Sequeira, C.A.C. Carbon Anode in Carbon History. *Molecules* **2020**, *25*, 4996. [CrossRef]
47. Tang, Y.; Liu, J. Effect of anode and Boudouard reaction catalysts on the performance of direct carbon solid oxide fuel cells. *Int. J. Hydrogen Energy* **2010**, *35*, 11188–11193. [CrossRef]
48. Han, T.; Wu, Y.; Xie, Z.; Wang, Y.; Xie, Y.; Zhang, J.; Xiao, J.; Yu, F.; Yang, N. A novel Boudouard reaction catalyst derived from strontium slag for enhanced performance of direct carbon solid oxide fuel cells. *J. Alloys Compd.* **2022**, *895*, 162643. [CrossRef]
49. Yu, J.; Zhao, Y.; Li, Y. Utilization of corn cob biochar in a direct carbon fuel cell. *J. Power Source* **2014**, *270*, 312–317. [CrossRef]
50. Technical Association of the Pulp and Paper Industry. *TAPPI Test Methods*; TAPPI: Atlanta, GA, USA, 2012.
51. Browning, B. *Methods of Wood Chemistry*; Interscience Publishers: New York, NY, USA, 1967.
52. Perry, R.H.; Chilton, C.H. *Chemical Engineers' Handbook*, 5th ed.; Cecil, H., Perry, J.H., Eds.; McGraw-Hill: New York, NY, USA, 1973.
53. Grioui, N.; Halouani, K.; Zoulalian, A.; Halouani, F. Experimental Study of Thermal Effect on Olive Wood Porous Structure During Carbonization. *Maderas. Cienc. Tecnol.* **2007**, *9*, 15–28. [CrossRef]
54. Liang, Y.; Cheng, B.; Si, Y.; Cao, D.; Jiang, H.; Han, G.; Liu, X. Thermal decomposition kinetics and characteristics of Spartina alterniflora via thermogravimetric analysis. *Renew. Energy* **2014**, *68*, 111–117. [CrossRef]
55. Ma, Z.; Chen, D.; Gu, J.; Bao, B.; Zhang, Q. Determination of pyrolysis characteristics and kinetics of palm kernel shell using TGA-FTIR and model-free integral methods. *Energy Convers. Manag.* **2015**, *89*, 251–259. [CrossRef]
56. García, G.B.; Calero De Hoces, M.; Martínez García, C.; Cotes Palomino, M.T.; Gálvez, A.R.; Martín-Lara, M.Á. Characterization and modeling of pyrolysis of the two-phase olive mill solid waste. *Fuel Process. Technol.* **2014**, *126*, 104–111. [CrossRef]
57. Cooper, J.F. Direct Conversion of Coal and Coal-Derived Carbon in Fuel Cells. In Proceedings of the 2nd International Conference on Fuel Cell Science, Engineering and Technology, Rochester, NY, USA, 14–16 June 2004; ASME: New York, NY, USA, 2004; pp. 375–385.
58. De Filippis, P.; Di Palma, L.; Petrucci, E.; Scarsella, M.; Verdone, N. Production and Characterization of Adsorbent Materials from Sewage Sludge by Pyrolysis. *Chem. Eng. Trans.* **2013**, *32*, 205–210.
59. Choudhury, N.D.; Chutia, R.S.; Bhaskar, T.; Kataki, R. Pyrolysis of jute dust: Effect of reaction parameters and analysis of products. *J. Mater. Cycles Waste Manag.* **2014**, *16*, 449–459. [CrossRef]
60. Di Blasi, C. Heat, Momentum and mass transport through a shrinking biomass particle exposed to thermal radiation. *Chem. Eng. Sci.* **1996**, *51*, 1121–1132. [CrossRef]
61. Rady, A.C.; Giddey, S.; Badwal, S.P.S.; Ladewig, B.P.; Bhattacharya, S. Review of fuels for direct carbon fuel cells. *Energy Fuels* **2012**, *26*, 1471–1488. [CrossRef]
62. Vutetakis, D.G.; Skidmore, D.R.; Byker, H.J. Electrochemical Oxidation of Molten Carbonate-Coal Slurries. *J. Electrochem. Soc.* **1987**, *134*, 3027–3035. [CrossRef]
63. Joyce, J.; Dixon, T.; Diniz Da Costa, J.C. Characterization of sugarcane waste biomass derived chars from pressurized gasification. *Process Saf. Environ. Prot.* **2006**, *84*, 429–439. [CrossRef]
64. Nzihou, A.; Stanmore, B.; Sharrock, P. A review of catalysts for the gasification of biomass char, with some reference to coal. *Energy* **2013**, *58*, 305–317. [CrossRef]
65. Mahadevan, R.; Adhikari, S.; Shakya, R.; Wang, K.; Dayton, D.; Lehrich, M.; Taylor, S.E. Effect of Alkali and Alkaline EarthMetals on in-Situ Catalytic Fast Pyrolysis of Lignocellulosic Biomass: A Micro-reactor Study. *Energy Fuels* **2016**, *30*, 3045–3056. [CrossRef]
66. Bourke, J.; Manley-Harris, M.; Fushimi, C.; Dowaki, K.; Nunoura, T.; Antal, M.J. Do all carbonized charcoals have the same chemical structure? *Ind. Eng. Chem. Res.* **2007**, *46*, 5954–5967. [CrossRef]
67. Keiluweit, M.; Nico, P.S.; Johnson, M.G.; Kleber, M. Dynamic Molecular Structure of Plant Biomass-derived Black Carbon (Biochar)-Supporting Information. *Environ. Sci. Technol.* **2010**, *44*, 1247–1253. [CrossRef] [PubMed]
68. Yuan, J.H.; Xu, R.K.; Zhang, H. The forms of alkalis in the biochar produced from crop residues at different temperatures. *Bioresour. Technol.* **2011**, *102*, 3488–3497. [CrossRef] [PubMed]
69. Ahmad, M.; Ahmad, M.; Usman, A.R.A.; Al-Faraj, A.S.; Abduljabbar, A.; Ok, Y.S.; Al-Wabel, M.I. Date palm waste-derived biochar composites with silica and zeolite: Synthesis, characterization and implication for carbon stability and recalcitrant potential. *Environ. Geochem. Health* **2017**, *41*, 1687–1704. [CrossRef]
70. Rafiq, M.K.; Bachmann, R.T.; Rafiq, M.T.; Shang, Z.; Joseph, S.; Long, R.L. Influence of pyrolysis temperature on physico-chemical properties of corn stover (*zeamays* L.) biochar and feasibility for carbon capture and energy balance. *PLoS ONE* **2016**, *11*, e0156894. [CrossRef] [PubMed]

71. Sigmund, G.; Hüffer, T.; Hofmann, T.; Kah, M. Biochar total surface area and total pore volume determined by $N_2$ and $CO_2$ physisorption are strongly influenced by degassing temperature. *Sci. Total Environ.* **2017**, *580*, 770–775. [CrossRef]
72. Li, X.; Zhu, Z.; DeMarco, R.; Bradley, J.; Dicks, A. Modification of coal as a fuel for the direct carbon fuel cell. *J. Phys. Chem. A* **2010**, *114*, 3855–3862. [CrossRef]
73. Li, J.; Wei, B.; Wang, C.; Zhou, Z.; Lu, Z. High performance and stable $La_{0.8}Sr_{0.2}Fe_{0.9}Nb_{0.1}O_{3-\delta}$ anode for direct carbon solid oxide fuel cells fueled by activated carbon and corn straw derived carbon. *Int. J. Hydrogen Energy* **2018**, *43*, 12358–12367. [CrossRef]
74. Wu, H.; Xiao, J.; Hao, S.; Yang, R.; Dong, P.; Han, L.; Li, M.; Yu, F.; Xie, Y.; Ding, J.; et al. In-situ catalytic gasification of kelp-derived biochar as a fuel for direct carbon solid oxide fuel cells. *J. Alloys Compd.* **2021**, *865*, 158922. [CrossRef]
75. Chien, A.C.; Arenillas, A.; Jiang, C.; Irvine, J.T.S. Performance of Direct Carbon Fuel Cells Operated on Coal and Effect of Operation Mode. *J. Electrochem. Soc.* **2014**, *161*, F588–F593. [CrossRef]
76. Mohan, D.; Pittman, C.U.; Steele, P.H. Pyrolysis of wood/biomass for bio-oil: A critical review. *Energy Fuels* **2006**, *20*, 848–889. [CrossRef]
77. Hoffmann, V.; Jung, D.; Zimmermann, J.; Correa, C.R.; Elleuch, A.; Halouani, K.; Kruse, A. Conductive carbon materials from the hydrothermal carbonization of vineyard residues for the application in electrochemical double-layer capacitors(EDLCs) and direct carbon fuel cells(DCFCs). *Materials* **2019**, *12*, 1703. [CrossRef] [PubMed]
78. Eom, S.; Ahn, S.; Rhie, Y.; Choi, G.; Kim, D. Effect of Coal Gases on Electrochemical Reactions in the Direct Carbon Fuel CellSystem. *J. Clean Energy Technol.* **2015**, *3*, 72–77. [CrossRef]
79. Adinaveen, T.; Vijaya, J.J.; Kennedy, L.J. Comparative Study of Electrical Conductivity on Activated Carbons Prepared from Various Cellulose Materials. *Arab. J. Sci. Eng.* **2016**, *41*, 55–65. [CrossRef]

**Disclaimer/Publisher's Note:** The statements, opinions and data contained in all publications are solely those of the individual author(s) and contributor(s) and not of MDPI and/or the editor(s). MDPI and/or the editor(s) disclaim responsibility for any injury to people or property resulting from any ideas, methods, instructions or products referred to in the content.

Article

# Pre-Feasibility Study of a Multi-Product Biorefinery for the Production of Essential Oils and Biomethane

Luís Carmo-Calado [1], Roberta Mota-Panizio [1,*], Ana Carolina Assis [1], Catarina Nobre [1], Octávio Alves [2], Gonçalo Lourinho [2] and Paulo Brito [1]

[1] VALORIZA—Research Centre for Endogenous Resource Valorization, Polytechnic Institute of Portalegre, 7300-555 Portalegre, Portugal
[2] CoLAB BIOREF—Collaborative Laboratory for Biorefineries, 4466-901 São Mamede de Infesta, Portugal
* Correspondence: rpanizio@ipportalegre.pt

**Abstract:** Rural areas can benefit from the development of biorefineries for the valorization of endogenous feedstocks. In this study, a pre-feasibility assessment of an integrated multi-product biorefinery to produce essential oils and biomethane is carried out considering current technical and economic conditions. The proposed concept is based on the steam distillation of forestry biomass for the extraction of essential oils (2900 L/y) followed by biomethane production via syngas methanation using the spent biomass as feedstock (30.4 kg/h). In parallel, the anaerobic treatment of WWTP sludge (5.3 kg/h) is used to produce additional biomethane for mobile applications. The results show that the intended multi-product biorefinery delivers attractive benefits for investors as described by the calculated financial indicators: NPV of EUR 4342.6, IRR of 18.1%, and PB of 6 years. Overall, the pre-feasibility analysis performed in this study demonstrates that the proposed biorefinery concept is promising and warrants further investment consideration via cost and benefit analysis, ultimately promoting the implementation of multi-product biorefineries across Europe.

**Keywords:** biorefineries; gasification; essential oils; biomass wastes; biomethane

## 1. Introduction

Biorefineries have been suggested to decrease the environmental and social issues caused by fossil resources by replacing fossil feedstocks with biological resources. In these infrastructures, biomass is fractionated into a multitude of value-added products and energy vectors capable of sustainably satisfying the energy and material needs of several industry sectors [1]. To achieve this wide range of products, biorefineries, such as conventional oil refineries, require the integration of different processes and technologies in a single facility, preferably.

Regarding feedstock for biorefineries, there have been many studies testing forestry biomass wastes, agricultural wastes, sludges from various sources, or municipal solid wastes (MSWs) [2–4]. Forestry biomass residues have received greater attention in this application, mostly as a response to the increasing global energy demand but also for their potential in the reduction in greenhouse gas (GHG) emissions [2]. These biomass wastes are renewable energy sources, and they are perceived as recycling carbon instead of removing it from long-term storage [3]. Another very promising feedstock for biorefineries is sludge, particularly sludges from wastewater treatment facilities (WWTPs). These materials are solid waste residues rich in organic compounds such as cellulose, which can represent approximately 20–50% of the influent suspended solids in WWTPs [5,6].

Because of its location and climate, Portugal is well-suited to forest growth, which covers about 35% of the territory. In this context, forestry wastes are a potential renewable feedstock for the country [7,8]. WWTP sludges are also very representative, constituting another potential waste to be used in biorefineries. For example, according to Santos et al. (2022), these sludges can be considered a valuable material source after proper treatment, contributing to the sustainable circular economy of the wastewater treatment sector [9]. Overall, several industries are producing very significant amounts of waste with good biorefining potentials such as food, chemical, textile, paints, resins, pharmaceuticals, tanneries, paper, metallurgy, and mining [9].

Processing biomass and wastes in biorefineries may require the integration of several technological processes, such as separation processes, chemical or biochemical conversions, and thermochemical conversions. Thus, biorefineries can be classified according to the type of technological process involved and defined in different platforms: biochemistry, thermochemistry, biodiesel, and biogas. The thermochemical platform involves the decomposition of biomass via gasification or pyrolysis, using heat and catalysts. Current developments require the improvement of thermochemical processes to higher operation efficiency, advancements in new equipment, and coupling with other technologies, such as electrolysis, methanation, or anaerobic digestion (AD), to expand the biomass feedstocks that can be used and the array of end products. With this more complex approach, also known as multi-product biorefineries, these infrastructures can yield energy, biofuels, and added-value products. One example of a multi-product biorefinery is the extraction of essential oils (EOs) from forestry biomass and the use of waste biomass from the process to produce biomethane via gasification and syngas methanation. In parallel, it is also possible to use other feedstocks in an anaerobic digestor to produce and upgrade biogas into biomethane, enhancing renewable gas production. The merged biomethane flows may then be used in mobility applications or for heat and electricity production. Consequently, this conceptual biorefinery concept based on technologically mature technologies would yield several marketable products, a low amount of generated waste, and improved yields.

EOs are one of the most interesting products that can be obtained in biorefineries using forestry biomass wastes. These compounds have been thoroughly studied throughout the years due to several pharmacological properties given by their main bioactive compounds (e.g., isoprenoids) [10]. In addition, EOs also present antimicrobial, antioxidant and anti-inflammatory properties, which explain the considerable interest in their extraction, as described by several authors [11–17]. Due to their features, EOs extracted from different feedstocks are commercialized and used in many applications such as food packaging, edible films and coatings [18–23], microencapsulation [24], biomedicine applications [25,26], and agricultural applications [27–30]. The high market value of essential oils could enable the use of waste forestry resources to be economically viable.

Usually, EOs are extracted by cold pressing, steam distillation—SD (which includes dry steam, direct steam, and hydro distillation), solvent-assisted extraction, ultrasonic-assisted extraction, supercritical fluid extraction, or solvent-free microwave extraction [31]. SD is the most conventionally used technique for EO extraction, albeit presenting lower yield and efficiency and higher extraction time than the other referred methods. Furthermore, SD has low capital and operational costs, making this technique very interesting for biorefinery integration [31,32]. Kant and Kumar (2022) analyzed conventional EO extraction techniques from rosemary and oregano and determined that production costs for EO extraction using SD varied between 14.90 and 71.93 EUR/kg [31]. EOs from rosemary and oregano were also studied by Moncada et al. (2016). The authors used water distillation (conventional) and supercritical fluid extraction (non-conventional) and concluded that energy integration played a relevant role in the pricing of EOs. Oregano EOs showed the lowest production costs by using supercritical fluid extraction with full energy integration (6.31 EUR/kg), while rosemary EOs had lower production using water distillation with full energy integration (6.18 EUR/kg) [33].

Gasification is the conversion of organic or carbonaceous raw materials at high temperatures. The process mainly produces gaseous products, including hydrogen ($H_2$), carbon monoxide (CO), small amounts of carbon dioxide ($CO_2$), nitrogen ($N_2$), water ($H_2O$), and hydrocarbons ($C_nH_m$) [34,35]. Biomass gasification is an old and economical alternative for the production of renewable gases. For example, the production of hydrogen can be achieved by the partial oxidation of wood particles using oxygen as the gasifying agent, yielding a hydrogen fraction directly in the syngas, which can be enhanced through the water–gas shift (WGS) reaction [36]. Low-temperature catalytic gasification is also an interesting alternative for hydrogen production from an energy point of view, as it requires a relatively low heat input, and gas treatment is not necessary. Both from an input–output point of view and the complexity involved in the process, low-temperature catalytic gasification becomes more attractive and viable than high-temperature gasification [37]. Furthermore, several processes are used to clean and condition the syngas to the quality needed, not only for hydrogen production but also for further chemical synthesis. Mature technologies (commercially available for syngas cleaning and upgrading) include the above-mentioned WGS reaction, scrubbers, membrane separation, or pressure swing adsorption (PSA).

AD is the current technological benchmark for biomethane production. The process uses microorganisms to convert organic compounds such as carbohydrates, proteins, and lipids into methane, carbon dioxide, water, and other vestigial compounds. AD is a well-established and mature technology used to treat sludges and other organic effluents [38,39]. Biogas, the main product resulting from the process, has enough methane content to contribute as a renewable energy vector; simultaneously, digestate can be used as a fertilizer due to its high nutrient concentration (N and P) [39]. Methanation, on the other hand, has also been receiving a lot of attention as a thermochemical pathway for biomethane production. Two main reactor concepts represent the state of the art in methanation technologies: adiabatic or cooled fixed-bed reactors and fluidized bed reactors. Adiabatic fixed beds are commercially available but typically increase the complexity of the process setup due to their inherent heat vulnerability. On the other hand, fluidized beds can avoid localized hot spot formation and increase the tolerance to unsaturated hydrocarbon traces in the feed gas, although they still lack technological maturity.

Despite their great potential to be a common point between different productive chains and industrial processing lines, biorefineries have not been widely implemented worldwide [40]. This is evident when collecting information on techno-economic analysis for multi-product bio-refineries. There is still a shortage of information regarding the costs involved in the implementation of biorefineries, more so when considering multiple technologies and multiple products. Despite this, some studies share relevance with the present work [2,41]. Michailos et al. (2020), for example, evaluated the techno-economic performance of a Power-to-Gas (P2G) system which closes the energy and material loops of an AD plant and produces high-purity methane from sewage sludge in a real wastewater plant (WWTP). The authors considered four production scenarios: biomethanation, biomethanation + gasification of the digestate for hydrogen production, biomethanatiom (with increased hydrogen and carbon dioxide) + gasification of the digestate for hydrogen production, and biomethanation + gasification of the digestate + integrated gasification combined cycle. The energy efficiency of the proposed concepts was found to be between 26.5% and 35.5%, with a minimum selling price (MSP) for biomethane between 154.8 and 209.8 EUR/MWh, with the possibility of being reduced by 34–42% with the implementation of some process improvements and by considering revenues from the process's by-products [41].

In this paper, the pre-feasibility of an integrated multi-product biorefinery yielding EOs and biomethane as major products is assessed. The concept involves the use of SD to fractionate mixtures of forestry biomass (mainly *E. globulus* and *C. ladanifer*) and the gasification of the resulting biomass to obtain syngas. This syngas is further cleaned and processed via catalytic methanation to obtain biomethane, while in parallel, an anaerobic digestor processes WWTP sludge to produce additional biomethane after biogas upgrading. The final biomethane uses considered in the study are mobility (e.g., heavy freight transportation) and heat and electricity production (e.g., solid oxide fuel cells).

## 2. Proposed Biorefinery Concept

The biorefinery concept considered in this work starts with the forest management practices from which biomass wastes are produced: pre-cleaning, cleaning, classification, transport, and final separation via particle dimension. After collection, part of the biomass, namely eucalyptus (*E. globulus*) and rockrose (*C. ladanifer*), is subjected to a steam distillation process in a 200 kg/h reactor to extract the EOs. These forest species were chosen considering their abundance in Portugal and the strong potential to become feedstocks in a biorefinery for the production of multiple products [42]. Steam for the SD process is obtained from the thermal energy produced in the gasification reactor. The spent biomass wastes from the extraction of EOs are then grounded and pelletized for subsequent gasification in a 1000 kg/h fluidized-bed gasifier at 800–95 °C. After gasification, the producer gas is cleaned through a cyclone filter and condenser, yielding char and ash (for soil applications) and condensates (which will be further introduced into the AD process). This gasifier has the particularity of operating with 50 vol.% oxygen produced by an electrolyzer (23.4 kg/h of hydrogen) coupled with photovoltaic panels. Finally, in the methanation reactor (fixed bed), carbon dioxide from the burning of the syngas is mixed with this green hydrogen and transformed into methane (30.42 kg/h).

In parallel, an AD reactor is fed with WWTP sludge, achieving a biogas production rate of 5.3 kg/h. WWTP sludge is also an extremely abundant and under-valorized waste in Portugal. Biogas upgrading proceeds through PSA, and the biomethane produced in the two technological pathways are combined and used for mobile applications or in an SOFC for the production of thermal and electrical energy. The following flowchart presents the multi-product biorefinery described above (Figure 1):

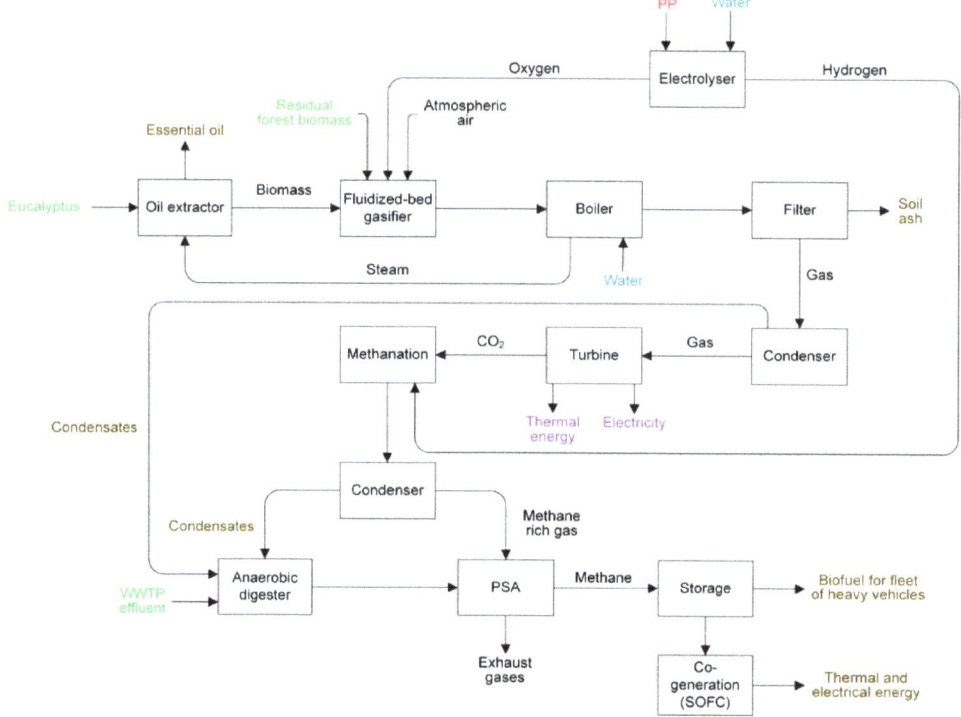

**Figure 1.** Representative flowchart of the proposed biorefinery concept: complex multi-product biorefinery.

## 3. Economic Analysis of the Proposed Biorefinery Concept

### 3.1. Assumptions for the Different Biorefinery Technologies

The biorefinery concept presented in this study aims to produce EOs and biomethane as the main products. The main goal is to assess the basic conceptual, economic, and financial viability of the concept and to identify the main costs and benefits of the proposed multi-product biorefinery.

Considering the details given in Section 2, the initial assumptions for the pre-feasibility assessment are presented in Table 1.

**Table 1.** Initial assumptions for the proposed complex multi-product biorefinery.

| Assumptions | Units | Value |
|---|---|---|
| Consumption of raw materials (gasification) | kg/h | 1000 |
| Percentage of pure $O_2$ (gasification) | % | 50 |
| $O_2$ flow rate (electrolysis) | m$^3$/h | 212.8 |
| Annual working hours | h | 7200 |
| Consumption of raw materials (essential oils) | kg/h | 200 |
| **Operational data** | **Units** | **Value** |
| Oxygen density | kg/m$^3$ | 1.43 |
| Percentage of $O_2$ in the air | % | 21 |
| Air flow | m$^3$/h | 1418 |
| $O_2$ flow rate in the air | m$^3$/h | 297.8 |
| Air flow for 100 kg of raw material | m$^3$/h | 141.8 |

More detailed data for each technological stage included in the biorefinery concept are presented below.

#### 3.1.1. Gasification Unit

For the gasification stage, a fluidized-bed gasification unit with a consumption capacity of 1000 kg/h was studied. The main gasification parameters considered in the study are shown in Table 2.

**Table 2.** Initial assumptions for the gasification unit.

| Parameters | Units | Value | Reference |
|---|---|---|---|
| Consumption of raw materials | kg/h | 1000 | — |
| Operating hours | h/year | 7200 | — |
| Gasifier cost | EUR/kWe | 5474 | [43] |
| Capital cost (5474 EUR/kWe) | EUR | 5,540,985.8 | [44] |
| Life cycle | years | 10 | — |
| Energy consumption | kWh | 300 | — |
| Start operation | year | 2021 | — |
| Raw material | EUR/t | 7.50 | — |
| Raw material costs | EUR/year | 54,000 | — |
| Operating costs | EUR/month | 18,075 | — |
| Maintenance costs (2%) | EUR/year | 110,819 | — |
| Thermal energy (77% of Ee and less Et boiler) | GWh/year | 5.57 | — |
| Electric energy sold (self-consumption and losses 10%) | GWh/year | 7.29 | — |
| Sales of electric power | EUR/kW | 0.12 | — |
| Sales of thermal energy | EUR/kW | 010 | — |

Note: Ee—electrical energy; Et—thermal energy.

#### 3.1.2. Electrolyzer

The electrolyzer was sized according to the oxygen requirements of the gasification unit (50 vol.% of oxidizing agent). The unit is powered by photovoltaic panels (PV electrolysis) and achieves a green hydrogen production rate of approximately 23 kg of $H_2$ per hour. Specific data related to the electrolysis unit are presented in Table 3.

Table 3. Initial assumptions for the electrolysis unit (hydrogen and oxygen production).

|  | Parameters | Units | Value | Reference |
|---|---|---|---|---|
|  | Capital cost | EUR/kW | 659 | [45] |
|  | Operating costs (2.2%) | EUR/kW/year | 14.5 | — |
|  | Water costs | EUR/m$^3$ | 1.5 | [46] |
|  | Use of water | L/kg of $H_2$ | 10 | — |
|  | Use of water | L/kg of $O_2$ | 1.10 | — |
|  | Efficiency | kWh/kg of $H_2$ | 54 | — |
|  | Maintenance costs | %/year | 0.30 | — |
|  | Battery life | h | 80,000 | — |
|  | Battery replacement costs (5.1%) | EUR | 36.6 | — |
| Photovoltaic | Capital cost | EUR/kW | 687 | — |
|  | Operating costs (1.20%) | EUR/kW/year | 8.25 | — |
|  | Annual degradation | %/year | 0.50 | — |
|  | Solar hours | h/year | 2500 | — |
| General | Years of operation | year | 10 | — |

3.1.3. Essential Oils Extraction Unit

The EOs extraction unit is an SD extraction unit, which has the particularity of recovering the thermal energy produced in gasification to generate the necessary steam for the extraction of EOs. A feedstock consumption of 200 kg/h yields the production of approximately 2900 L of EOs per year. Table 4 presents the parameters defined for this unit.

Table 4. Initial assumptions for the essential oil extraction unit.

| Parameters | Units | Value | Reference |
|---|---|---|---|
| Water consumption | L/h | 100 | — |
| Consumption of raw materials | kg/h | 200 | — |
| Operating hours | h/year | 7200 | — |
| Water costs | EUR/m$^3$ | 1.5 | — |
| Use of water | L/kg of EO | 250 | — |
| Ee consumption | kWh | 20 | — |
| Eth consumption | kWth | 6 | — |
| Start operation | year | 2021 | — |
| Raw material costs | EUR/t | 7.5 | — |
| OE value (*E.globulus*) | EUR/kg | 20 | [47,48] |
| OE sales | EUR/day | 469.3 | — |
| Years of operation | Year | 10 | — |

3.1.4. Methanation Unit

The methanation unit is interconnected with the exhaust gases produced during the gasification process and the hydrogen produced from electrolysis. The methane produced, about 30 kg/h, will be used for mobility. The data related to this unit are presented in Table 5.

**Table 5.** Initial assumptions for the methanation unit.

| Parameters | Units | Value | Reference |
|---|---|---|---|
| Hours of operation | h | 7200 | — |
| Capital cost | EUR/kW | 400 | [49,50] |
| Total cost | EUR | 27,185.6 | — |
| Efficiency | kWh/kg of $CH_4$ | 2.2 | [51] |
| Amount of $CH_4$ | kg/h | 30.4 | — |
| Price of $CH_4$ | EUR/kg | 0.97 | [52] |
| Annual methane sales | EUR/year | 212,489.8 | — |
| Energy quantity | kWh | 68 | — |
| Maintenance costs (3%/year) | EUR | 815.6 | — |

3.1.5. Anaerobic Digestion Unit

The anaerobic digestion unit is characterized by the degradation of the organic fraction of sludge (11 m$^3$/day) while simultaneously producing biogas. The unit will have a biomethane production capacity of around 190 m$^3$/day. The data relating to this unit are presented in Table 6.

**Table 6.** Initial assumptions for the anaerobic digestion unit.

| Parameters | Units | Value |
|---|---|---|
| Consumption of raw materials | m$^3$/day | 11 |
| Operating hours | h/year | 8760 |
| Capital cost | EUR | 16,740.9 |
| Life cycle | Years | 10 |
| Energy consumption | kWh | 15 |
| Volume | m$^3$ | 220 |
| Daily biogas production | m$^3$/day | 275 |
| Daily biomethane production | m$^3$/day | 192.5 |
| Daily biomethane production | kg/day | 126.5 |
| Daily biomethane production | kg/h | 5.3 |
| Price of $CH_4$ | EUR/kg | 0.97 |
| Annual methane sales | EUR/year | 44,777.6 |

3.1.6. Pressure Swing Adsorption (PSA) Unit

PSA is a technique used to separate gaseous compounds from a mixture of gases under pressure, according to the molecular characteristics of the species and affinity for an adsorbing material. The parameters related to PSA can be found in Table 7.

**Table 7.** Initial assumptions for the PSA unit.

| Parameters | Units | Value |
|---|---|---|
| Hours of operation | h | 7200 |
| Capital cost | EUR/kW | 1000 |
| Total cost | EUR | 30,000 |
| Energy quantity | kWh | 30 |

3.1.7. Solid Oxide Fuel Cell (SOFC) Unit

An SOFC produces electrical and thermal energy through the oxidation of a fuel, in this case, methane. The cell will oxidize about 30 kg/h of methane and produce about 100 kWh. All data relating to the SOFC unit are described in Table 8.

Table 8. Initial assumptions for the SOFC unit.

| Parameters | Units | Value | Reference |
|---|---|---|---|
| Hours of operation | h | 7200 | — |
| Capital cost | EUR/kW | 1500 | — |
| Total cost | EUR | 1,349,108 | — |
| Electrical efficiency | $kW_{eh}$ | 89.9 | [53] |
| Thermal efficiency | $kW_{thh}$ | 86.2 | — |
| Amount of $CH_4$ | kg/h | 30 | — |
| Consumption cost $CH_4$ | EUR/h | 29.1 | — |

## 4. Results and Discussion

Tables 9–12 detail the economic assumptions used to build the pre-feasibility model of the proposed biorefinery concept. The assessment is based on the discounted cash flow from which the actual feasibility of the project can be inferred through the calculation of the net present value (NPV) of the project, as well as the internal rate of return (IRR) and the payback period (PP). These three economic parameters are common indicators in investment decisions. In particular, the NPV yields the current value of the investment project, as well as its profitability, by updating the entire cash flow of an investment to its present value using a proper discount rate based on macroeconomic conditions. For the NPV, it is stated that an investment should be accepted if the NPV >0 and rejected if the NPV <0. IRR is obtained by calculating the discount rate that produces an NPV equal to zero, whereas the payback is defined as the minimum period (in years) needed to recover the initial capital investments made, i.e., the year in which the cumulative cash flows become positive.

Table 9. Initial investment for the proposed multi-product biorefinery.

|  | Parameters | Units | Value |
|---|---|---|---|
| Electrolyzer |  |  | 1,701,097.7 |
| EOs extractor |  |  | 40,000 |
| Gasifier |  |  | 5,540,985.8 |
| Methanation | Capital cost | EUR | 27,185,6 |
| DA |  |  | 16,740,9 |
| PSA |  |  | 30,000 |
| SOFC |  |  | 134,910.8 |
|  |  | TOTAL | EUR 7,490,920.7 |

Table 10. Annual consumption in kWh of the units that make up the proposed biorefinery concept.

|  | Parameters | Units | Nominal |
|---|---|---|---|
| Electrolyzer |  |  | - |
| EO extractor |  |  | 20 |
| Gasifier |  |  | 200 |
| Methanation | Consumption | kWh | 200 |
| DA |  |  | 15 |
| PSA |  |  | 30 |
| SOFC |  |  | 10 |
|  |  | TOTAL | 475 kWh |
|  |  | TOTAL ANNUAL | 3,466,800 kWh |

Table 11. Costs associated with maintenance, operation, and feedstock.

|  | Parameters | Units | Value |
|---|---|---|---|
| Operation |  | EUR/year | 168,700 |
| Workers |  | n° | 10 |
| Wages |  | EUR/month | 1205 |
| Raw material |  |  | 54,000 |
| Maintenance | 1.5% year of total costs |  | 112,363.8 |
|  |  | TOTAL | EUR 335,063.8 |

Table 12. Revenues associated with the proposed biorefinery concept.

|  | Parameters | Units | Value |
|---|---|---|---|
| Sales of EOs |  |  | 422,400 |
| Vehicular biomethane | methanation + DA—SOFC |  | 56,811.2 |
| Purchase of electrical energy | self-consumption | EUR/year | 520,020 |
| Sale of electrical energy | 2.43 gasifier + 0.22 SOFC—self-consumption |  | 484,459.7 |
| Thermal energy | 1.85 gasifier + 0.21 SOFC |  | 577,553.3 |
|  |  | TOTAL | EUR 2,061,244.2 |

The first steps in the analysis comprised the estimation of benefits and costs for each process stage to determine overall cash inflows and outflows. The cash flows considered were the initial investment, operation, and maintenance costs and revenues from sales of electric energy (considering self-consumption), thermal energy, biomethane for mobility, and EOs. All cash flows, except for the initial investment that occurs only in the start-up phase of the project, extend over the 10 years of the project's life, with all costs and revenues updated for the corresponding year. The total annual cash flow is the sum of all costs and revenues for each year. The annual revenue is given by multiplying the annual electricity production by the electricity price and the corresponding savings in the purchase of electricity due to self-consumption, sales of thermal energy, sales of vehicular biomethane, and sales of EOs. Lastly, the cumulative NPV is determined to give the present value of negative and positive investment cash flows. All analyses were performed at current prices, revenues, and value-added tax rates. The inflation rates implemented for 2021 and 2022 are based on Bank of Portugal forecasts and did not consider the current inflation rate due to adverse economic conditions arising from the war in Ukraine and post-COVID constraints.

Figure 2 presents the cumulative cash flows associated with the project in current prices. The calculation of economic parameters was carried out using the discounted cash inflows and outflows estimated in the figure, comparing economic costs and benefits over the project lifetime using a discount rate of 5.75%.

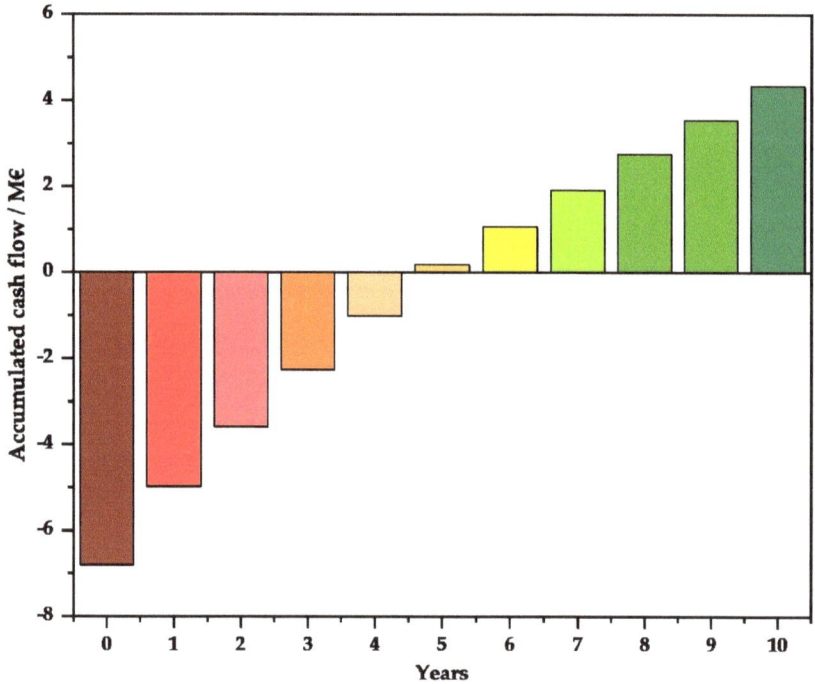

**Figure 2.** Cumulative cash flows for the multi-product biorefinery concept in current prices.

The proposed multi-product biorefinery presents an NPV of EUR 4342.6, an IRR of 18.1%, and a PB of 6 years. These results show that the project has a good chance of delivering positive economic benefits in the conditions studied. However, the analysis should go beyond the specific numbers, and the economic attractiveness of the project should be assessed using similar projects as baseline scenarios. In this case, direct comparison with other literature studies on EO extraction is difficult due to the novelty of the multi-product biorefinery presented here.

From an investor's point of view, a more general financial benchmark for biomass projects can be used for comparison: projects with NPVs higher than zero, IRRs greater than 10%, and PBs less than 10 years should advance from the pre-feasibility stage and assessment towards an investment decision should continue. Given these premises, it can be concluded that the pre-feasibility study of the biorefinery concept proposed in this work is promising in terms of its economic viability. Future studies may consider performing a comprehensive cost and benefit analysis and an overall assessment of the strategic, economic, and financial cases for the multi-product biorefinery concept studied. This analysis may include detailed market research and technical analysis, sustainability assessment, and investment appraisal regarding the implementation of this and other innovative concepts to enhance the value of endogenous resources.

## 5. Conclusions

The pre-feasibility of a multi-product biorefinery for the extraction of EOs and the production of biomethane was discussed and assessed considering current technical and economic conditions. In particular, the production of EOs (2900 L/y) was studied using the steam distillation of forestry biomass, while biomethane production was explored using the gasification of the spent biomass (30.4 kg/h) combined with the AD of WWTP sludge (5.3 kg/h) to maximize renewable gas production for different applications.

The pre-feasibility analysis showed that the intended multi-product biorefinery concept is promising and delivers positive economic benefits in the conditions studied. From the investor's perspective, results for the main financial indicators showed an NPV of EUR 4342.6, an IRR of 18.1%, and a PB of 6 years, values that are above current financial benchmarks in biomass projects.

Overall, this work demonstrated the conceptual viability of a multi-product biorefinery to produce EOs and biomethane from forestry wastes and sludge, therefore showing that the investment opportunity warrants further study. The next steps should include a comprehensive cost and benefit analysis, including detailed technical analysis, investment appraisal, and sustainability assessment, in order to ensure social and environmental benefits from the implementation of biorefineries in rural areas.

**Author Contributions:** Conceptualization, L.C.-C. and P.B.; Methodology, L.C.-C. and R.M.-P.; Validation, C.N.; Formal Analysis, C.N. and G.L.; Investigation, L.C.-C., A.C.A., R.M.-P., O.A. and G.L.; Writing—Draft Preparation, L.C.-C., R.M.-P. and A.C.A.; Writing—Review and Editing, C.N. and G.L.; Visualization, G.L. and O.A.; Supervision, P.B. All authors have read and agreed to the published version of the manuscript.

**Funding:** This research was funded by Fundação para a Ciência e a Tecnologia, I.P. (Portuguese Foundation for Science and Technology), under the project UIDB/05064/2020 (VALORIZA—Research Centre for Endogenous Resource Valorization), by the Regional Operational Program of Alentejo (Alentejo2020) under Portugal 2020 (Operational Program for Competitiveness and Internationalization) grant ALT20–05–3559–FSE– 000035, and by project PigWasteBiorefinery with grant ALT20-03-0246-FEDER-000054.

**Conflicts of Interest:** The authors declare no conflict of interest.

# References

1. Mariana, O.S.; Camilo, S.T.J.; Ariel, C.A.C. A Comprehensive Approach for Biorefineries Design Based on Experimental Data, Conceptual and Optimization Methodologies: The Orange Peel Waste Case. *Bioresour. Technol.* **2021**, *325*, 124682. [CrossRef] [PubMed]
2. Zhang, L.; Yang, P.; Zhu, K.; Ji, X.; Ma, J.; Mu, L.; Ullah, F.; Ouyang, W.; Li, A. Biorefinery-Oriented Full Utilization of Food Waste and Sewage Sludge by Integrating Anaerobic Digestion and Combustion: Synergistic Enhancement and Energy Evaluation. *J. Clean. Prod.* **2022**, *380*, 134925. [CrossRef]
3. Levasseur, A.; Bahn, O.; Beloin-Saint-Pierre, D.; Marinova, M.; Vaillancourt, K. Assessing Butanol from Integrated Forest Biorefinery: A Combined Techno-Economic and Life Cycle Approach. *Appl. Energy* **2017**, *198*, 440–452. [CrossRef]
4. Madadian, E.; Haelssig, J.B.; Mohebbi, M.; Pegg, M. From Biorefinery Landfills towards a Sustainable Circular Bioeconomy: A Techno-Economic and Environmental Analysis in Atlantic Canada. *J. Clean. Prod.* **2021**, *296*, 126590. [CrossRef]
5. Jing, H.; Wang, H.; Lin, C.S.K.; Zhuang, H.; To, M.H.; Leu, S.Y. Biorefinery Potential of Chemically Enhanced Primary Treatment Sewage Sludge to Representative Value-Added Chemicals - A de Novo Angle for Wastewater Treatment. *Bioresour. Technol.* **2021**, *339*, 125583. [CrossRef]
6. Edifor, S.Y.; Nguyen, Q.D.; van Eyk, P.; Biller, P.; Lewis, D.M. Rheological Studies of Municipal Sewage Sludge Slurries for Hydrothermal Liquefaction Biorefinery Applications. *Chem. Eng. Res. Des.* **2021**, *166*, 148–157. [CrossRef]
7. da Costa, T.P.; Quinteiro, P.; Arroja, L.; Dias, A.C. Environmental Comparison of Forest Biomass Residues Application in Portugal: Electricity, Heat and Biofuel. *Renew. Sustain. Energy Rev.* **2020**, *134*. [CrossRef]
8. Ferreira, S.; Monteiro, E.; Brito, P.; Vilarinho, C. Biomass Resources in Portugal: Current Status and Prospects. *Renew. Sustain. Energy Rev.* **2017**, *78*, 1221–1235. [CrossRef]
9. Santos, M.T.; Lopes, P.A. Sludge Recovery from Industrial Wastewater Treatment. *Sustain. Chem. Pharm.* **2022**, *29*, 1–18. [CrossRef]
10. Dajic Stevanovic, Z.; Sieniawska, E.; Glowniak, K.; Obradovic, N.; Pajic-Lijakovic, I. Natural Macromolecules as Carriers for Essential Oils: From Extraction to Biomedical Application. *Front. Bioeng. Biotechnol.* **2020**, *8*, 1–24. [CrossRef]
11. Hou, T.; Sana, S.S.; Li, H.; Xing, Y.; Nanda, A.; Netala, V.R.; Zhang, Z. Essential Oils and Its Antibacterial, Antifungal and Anti-Oxidant Activity Applications: A Review. *Food Biosci.* **2022**, *47*, 101716. [CrossRef]
12. Ilić, Z.S.; Milenković, L.; Tmušić, N.; Stanojević, L.; Stanojević, J.; Cvetković, D. Essential Oils Content, Composition and Antioxidant Activity of Lemon Balm, Mint and Sweet Basil from Serbia. *Lwt* **2022**, *153*, 112210. [CrossRef]
13. Gulluce, M.; Sahin, F.; Sokmen, M.; Ozer, H.; Daferera, D.; Sokmen, A.; Polissiou, M.; Adiguzel, A.; Ozkan, H. Antimicrobial and Antioxidant Properties of the Essential Oils and Methanol Extract from Mentha Longifolia L. Ssp. Longifolia. *Food Chem.* **2007**, *103*, 1449–1456. [CrossRef]
14. Pérez-Izquierdo, C.; Serrano-Pérez, P.; Rodríguez-Molina, M.d.C. Chemical Composition, Antifungal and Phytotoxic Activities of Cistus Ladanifer L. Essential Oil and Hydrolate. *Biocatal. Agric. Biotechnol.* **2022**, *45*, 102527. [CrossRef]

15. Benali, T.; Bouyahya, A.; Habbadi, K.; Zengin, G.; Khabbach, A.; Achbani, E.H.; Hammani, K. Chemical Composition and Antibacterial Activity of the Essential Oil and Extracts of Cistus Ladaniferus Subsp. Ladanifer and Mentha Suaveolens against Phytopathogenic Bacteria and Their Ecofriendly Management of Phytopathogenic Bacteria. *Biocatal. Agric. Biotechnol.* **2020**, *28*, 101696. [CrossRef]
16. Davari, M.; Ezazi, R. Chemical Composition and Antifungal Activity of the Essential Oil of Zhumeria Majdae, Heracleum Persicum and Eucalyptus Sp. against Some Important Phytopathogenic Fungi. *J. Mycol. Med.* **2017**, *27*, 463–468. [CrossRef]
17. Tavares, C.S.; Martins, A.; Faleiro, M.L.; Miguel, M.G.; Duarte, L.C.; Gameiro, J.A.; Roseiro, L.B.; Figueiredo, A.C. Bioproducts from Forest Biomass: Essential Oils and Hydrolates from Wastes of Cupressus Lusitanica Mill. and Cistus Ladanifer L. *Ind. Crops Prod.* **2020**, *144*, 112034. [CrossRef]
18. Korte, I.; Kreyenschmidt, J.; Wensing, J.; Bröring, S.; Frase, J.N.; Pude, R.; Konow, C.; Havelt, T.; Rumpf, J.; Schmitz, M.; et al. Can Sustainable Packaging Help to Reduce Food Waste ? A Status Quo Focusing Plant-Derived Polymers and Additives. *Appl. Sci.* **2021**, *11*, 5307. [CrossRef]
19. Perumal, A.B.; Huang, L.; Nambiar, R.B.; He, Y.; Li, X.; Sellamuthu, P.S. Application of Essential Oils in Packaging Films for the Preservation of Fruits and Vegetables: A Review. *Food Chem.* **2022**, *375*, 131810. [CrossRef]
20. Demircan, B.; Özdestan Ocak, Ö. The Effects of Ethyl Lauroyl Arginate and Lemon Essential Oil Added Edible Chitosan Film Coating on Biogenic Amines Formation during Storage in Mackerel Fillets. *J. Food Process. Preserv.* **2021**, *45*, 1–12. [CrossRef]
21. Al-Ali, R.M.; Al-Hilifi, S.A.; Rashed, M.M.A. Fabrication, Characterization, and Anti-free Radical Performance of Edible Packaging-chitosan Film Synthesized from Shrimp Shell Incorporated with Ginger Essential Oil. *J. Food Meas. Charact.* **2021**, *15*, 2951–2962. [CrossRef]
22. Varghese, S.A.; Siengchin, S.; Parameswaranpillai, J. Essential Oils as Antimicrobial Agents in Biopolymer-Based Food Packaging - A Comprehensive Review. *Food Biosci.* **2020**, *38*, 100785. [CrossRef]
23. Hafsa, J.; ali Smach, M.; Ben Khedher, M.R.; Charfeddine, B.; Limem, K.; Majdoub, H.; Rouatbi, S. Physical, Antioxidant and Antimicrobial Properties of Chitosan Films Containing Eucalyptus Globulus Essential Oil. *Lwt* **2016**, *68*, 356–364. [CrossRef]
24. Ibáñez, M.D.; Sanchez-Ballester, N.M.; Blázquez, M.A. Encapsulated Limonene: A Pleasant Lemon-like Aroma with Promising Application in the Agri-Food Industry. A Review. *Molecules* **2020**, *25*, 2598. [CrossRef]
25. Zakeri, Z.; Allafchian, A.; Vahabi, M.R.; Jalali, S.A.H. Synthesis and Characterization of Antibacterial Silver Nanoparticles Using Essential Oils of Crown Imperial Leaves, Bulbs and Petals. *Micro Nano Lett.* **2021**, *16*, 533–539. [CrossRef]
26. Pirsa, S. Evaluation of Release and Antibacterial Properties of Alginate Hydrogel Containing Beta-Cyclodextrin Nanoparticles. *Iran. J. Food Sci. Technol.* **2018**, 49–67. [CrossRef]
27. Isman, M.B.; Miresmailli, S.; MacHial, C. Commercial Opportunities for Pesticides Based on Plant Essential Oils in Agriculture, Industry and Consumer Products. *Phytochem. Rev.* **2011**, *10*, 197–204. [CrossRef]
28. Bohounton, R.B.; Sovegnon, P.M.; Barea, B.; Villeneuve, P. Aeollanthus Pubescens Benth Leaf Essential Oil: Its Chemical Composition and the Insecticidal Activity Against the Malaria Vector Anopheles Gambiae (Diptera : Culicidae). 2021, 1–16.
29. Srivastava, R. Insecticidal Activities of Some Essential Oils on Subterranean Termites. *Int. J. Innov. Sci. Res. Technol.* **2021**, *6*, 290–293.
30. Soares de Oliveira, M.A.; Melo Coutinho, H.D.; Jardelino de Lacerda Neto, L.; Castro de Oliveira, L.C.; Bezerra da Cunha, F.A. Repellent Activity of Essential Oils against Culicids: A Review. *Sustain. Chem. Pharm.* **2020**, *18*. [CrossRef]
31. Kant, R.; Kumar, A. Review on Essential Oil Extraction from Aromatic and Medicinal Plants: Techniques, Performance and Economic Analysis. *Sustain. Chem. Pharm.* **2022**, *30*, 100829. [CrossRef]
32. Machado, C.A.; Oliveira, F.O.; de Andrade, M.A.; Hodel, K.V.S.; Lepikson, H.; Machado, B.A.S. Steam Distillation for Essential Oil Extraction: An Evaluation of Technological Advances Based on an Analysis of Patent Documents. *Sustainability* **2022**, *14*, 7119. [CrossRef]
33. Moncada, J.; Tamayo, J.A.; Cardona, C.A. Techno-Economic and Environmental Assessment of Essential Oil Extraction from Oregano (Origanum Vulgare) and Rosemary (Rosmarinus Officinalis) in Colombia. *J. Clean. Prod.* **2016**, *112*, 172–181. [CrossRef]
34. Barneto, A.G.; Carmona, J.A.; Gálvez, A.; Conesa, J.A. Effects of the Composting and the Heating Rate on Biomass Gasification. *Energy and Fuels* **2009**, *23*, 951–957. [CrossRef]
35. Mota-Panizio, R.; Hermoso-Orzáez, M.J.; Carmo-Calado, L.; Calado, H.; Goncalves, M.M.; Brito, P. Co-Carbonization of a Mixture of Waste Insulation Electric Cables (WIEC) and Lignocellulosic Waste, for the Removal of Chlorine: Biochar Properties and Their Behaviors. *Fuel* **2022**, *320*. [CrossRef]
36. Manara, P.; Zabaniotou, A. Towards Sewage Sludge Based Biofuels via Thermochemical Conversion - A Review. *Renew. Sustain. Energy Rev.* **2012**, *16*, 2566–2582. [CrossRef]
37. Moghtaderi, B. Effects of Controlling Parameters on Production of Hydrogen by Catalytic Steam Gasification of Biomass at Low Temperatures. *Fuel* **2007**, *86*, 2422–2430. [CrossRef]
38. Edwards, J.; Othman, M.; Burn, S. A Review of Policy Drivers and Barriers for the Use of Anaerobic Digestion in Europe, the United States and Australia. *Renew. Sustain. Energy Rev.* **2015**, *52*, 815–828. [CrossRef]
39. Jiang, X.; Lyu, Q.; Bi, L.; Liu, Y.; Xie, Y.; Ji, G.; Huan, C.; Xu, L.; Yan, Z. Improvement of Sewage Sludge Anaerobic Digestion through Synergistic Effect Combined Trace Elements Enhancer with Enzyme Pretreatment and Microbial Community Response. *Chemosphere* **2022**, *286*, 131356. [CrossRef]
40. Solarte-Toro, J.C.; Laghezza, M.; Fiore, S.; Berruti, F.; Moustakas, K.; Cardona Alzate, C.A. Review of the Impact of Socio-Economic Conditions on the Development and Implementation of Biorefineries. *Fuel* **2022**, *328*, 125169. [CrossRef]

41. Michailos, S.; Walker, M.; Moody, A.; Poggio, D.; Pourkashanian, M. Biomethane Production Using an Integrated Anaerobic Digestion, Gasification and CO2 Biomethanation Process in a Real Waste Water Treatment Plant: A Techno-Economic Assessment. *Energy Convers. Manag.* **2020**, *209*, 112663. [CrossRef]
42. Mediavilla, I.; Guillamón, E.; Ruiz, A.; Esteban, L.S. Essential Oils from Residual Foliage of Forest Tree and Shrub Species: Yield and Antioxidant Capacity. *Molecules* **2021**, *26*, 3257. [CrossRef] [PubMed]
43. IRENA, I.R.E.A. RENEWABLE ENERGY TECHNOLOGIES: COST ANALYSIS SERIES. *IRENA Work. Pap. IRENA* **2012**, 1–60.
44. Kulkarni, A.; Baker, R.; Abdoulmomine, N.; Adhikari, S.; Bhavnani, S. Experimental Study of Torrefied Pine as a Gasification Fuel Using a Bubbling Fluidized Bed Gasifier. *Renew. Energy* **2016**, *93*, 460–468. [CrossRef]
45. Yates, J.; Daiyan, R.; Patterson, R.; Egan, R.; Amal, R.; Ho-Baille, A.; Chang, N.L. Techno-Economic Analysis of Hydrogen Electrolysis from Off-Grid Stand-Alone Photovoltaics Incorporating Uncertainty Analysis. *Cell Reports Phys. Sci.* **2020**, *1*, 100209. [CrossRef]
46. Lee, B.; Chae, H.; Choi, N.H.; Moon, C.; Moon, S.; Lim, H. Economic Evaluation with Sensitivity and Profitability Analysis for Hydrogen Production from Water Electrolysis in Korea. *Int. J. Hydrogen Energy* **2017**, *42*, 6462–6471. [CrossRef]
47. García, C.; Montero, G.; Coronado, M.A.; Valdez, B.; Stoytcheva, M.; Rosas, N.; Torres, R.; Sagaste, C.A. Valorization of Eucalyptus Leaves by Essential Oil Extraction as an Added Value Product in Mexico. *Waste Biomass Valorization* **2017**, *8*, 1187–1197. [CrossRef]
48. Mediavilla, I.; Blázquez, M.A.; Ruiz, A.; Esteban, L.S. Influence of the Storage of Cistus Ladanifer l. Bales from Mechanised Harvesting on the Essential Oil Yield and Qualitative Composition. *Molecules* **2021**, *26*, 2379. [CrossRef]
49. DVGW Der DVGW Sorgt Für Innovative Forschung Auf Höchstem Niveau.
50. Giuliano, A.; Freda, C.; Catizzone, E. Techno-Economic Assessment of Bio-Syngas Production for Methanol Synthesis: A Focus on the Water–Gas Shift and Carbon Capture Sections. *Bioengineering* **2020**, *7*, 70. [CrossRef]
51. Fernandez, B.F.; Menéndez, J.A. Syngas Production by CO2 Reforming of CH4 under Microwave Heating - Challenges and Opportunities. In *Syngas: Production, Applications and Environmental Impact*; Indarto, A., Palguandi, J., Eds.; Nova Science Publishers, Inc.: Hauppauge, NY, USA, 2011; pp. 121–149; ISBN 9781617617614.
52. Statista Average Price of Automotive Methane in Italy from 2010 to 2018 (in Euros Per Kilogram).
53. Gandiglio, M.; Drago, D.; Santarelli, M. Techno-Economic Analysis of a Solid Oxide Fuel Cell Installation in a Biogas Plant Fed by Agricultural Residues and Comparison with Alternative Biogas Exploitation Paths. *Energy Procedia* **2016**, *101*, 1002–1009. [CrossRef]

**Disclaimer/Publisher's Note:** The statements, opinions and data contained in all publications are solely those of the individual author(s) and contributor(s) and not of MDPI and/or the editor(s). MDPI and/or the editor(s) disclaim responsibility for any injury to people or property resulting from any ideas, methods, instructions or products referred to in the content.

*Article*

# Development of the Correlation Model between Biogas Yield and Types of Organic Mass and Analysis of Its Key Factors

Tetiana Mirzoieva [1], Nazar Tkach [1], Vitalii Nitsenko [2,3,*], Nataliia Gerasymchuk [4], Olga Tomashevska [5] and Oleksandr Nechyporenko [6]

1. Department of Economic, National University of Life and Environmental Sciences of Ukraine, 03041 Kyiv, Ukraine
2. Department of Entrepreneurship and Marketing, Institute of Economics and Management, Ivano-Frankivsk National Technical Oil and Gas University, 76019 Ivano-Frankivsk, Ukraine
3. SCIRE Foundation, 00867 Warsaw, Poland
4. Department of Cosmetology, University of Humanities and Economics in Lodz, 90-212 Lodz, Poland
5. Department of Entrepreneurship Organization and Exchange Activities, National University of Life and Environmental Sciences of Ukraine, 03041 Kyiv, Ukraine
6. Department of Land Relations and Environmental Management, National Scientific Centre "Institute of Agrarian Economics", 03127 Kyiv, Ukraine
* Correspondence: vitaliinitsenko@onu.edu.ua; Tel.: +380-939983073

**Abstract:** Since European society is experiencing an aggravation of the issue of energy security, the production of renewable energy is becoming increasingly important. The advantages of biofuel—in particular, biogas—and the positive effects of the development of its production are summarized within the framework of the problem statement. It is emphasized that the production of biogas from various renewable raw materials causes economic, ecological, and social effects. The development of biogas production can be especially active in combination with the development of the agricultural sphere. In response to today's demand, the authors in this research present a model of the correlation between the output of biogas from different types of organic mass and specify the factors affecting it. In particular, a multiple econometric model of the relationship between the output of biogas from different types of organic mass and the content of dry organic matter and the share of possible methane content in organic matter was built; the density of the connection between the factors and the resulting feature was evaluated; the tightness of the general relationship (influence) of independent variables on the dependent variable was checked using the coefficient of determination; and the reliability of the correlation characteristics was estimated using Fisher's and Student's tests. As a result, with the use of convincing evidence—in particular, taking into account the potential of the Ukrainian agricultural sector—the feasibility of further development of biogas production in combination with the development of agricultural production is substantiated.

**Keywords:** biogas; biogas production; bioenergy; energy security; organic mass

## 1. Introduction/Literature Review

In today's conditions, humanity is losing biodiversity on the planet, living in an era of large-scale environmental pollution, desertification, intensification of negative weather phenomena, loss and pollution of fresh water, and loss of forest resources. Stated above are either results of climate change or causes. It is a well-known fact that humanity has until 2030, when climate change will become irreversible. Against the background of such threatening prospects, as well as due to the depletion of fossil fuels, increasing global energy demand, ever-increasing fuel prices, and military conflicts that disrupt the energy balance of nations, humanity is increasingly focusing on alternative energy sources and efficient renewable energy sources.

Accordingly, in order to find an alternative to fossil fuels, as well as to solve the problem of landfilling of organic waste and reduce the negative impact on the environment in general, the world has been developing innovative approaches to bioenergy, biomaterials, and chemical production from organic waste for a long time. One of the key alternatives to fossil fuels is biofuels. Biomass is a potential renewable source for the production of solid, liquid, and gaseous biofuels. In turn, the modernization of existing technologies for the processing of organic waste into biofuels and its use as a substitute for natural gas or automotive fuel is a trendy direction in research in recent decades around the world. It is widely believed that the most viable option for obtaining energy from organic waste is biogas—a renewable gaseous fuel obtained by the decomposition of organic substances, such as food and animal husbandry waste [1,2].

Biogas consists mainly of methane and carbon dioxide. Combustible methane is the main component of biomass (50–85%), which is the main source of energy. It can be used in many ways, including automotive fuel, as well as for heating and electricity generation. In addition, the production of biogas produces a by-product—digestate, which can be used as biofertilizer. Digestate—or as it is also called, natural fertilizer—contains water, nutrients, and organic carbon suitable for soils [3–5].

The website of the European Biogas Association states that the use of digestate obtained in the process of biogas production as a biofertilizer helps return organic carbon back to the soil and reduces the need for carbon-laden mineral fertilizers [6].

We consider that information about this fact and everything related to biofuel production should continue to be actively disseminated and promoted both in society as a whole and among the farming environment, especially in developing countries.

It should be noted that the production and use of biogas is not an achievement of today; it has deep roots. It is believed that the first human use of biogas dates back to 3000 BC in the Middle East, when the Assyrians used biogas to heat their baths. In turn, the 17th century chemist Jan Baptist van Helmont discovered that combustible gases can be formed from decomposing organic substances. Van Helmont also first added the word «gas» to the scientific dictionary, derived from the Greek word «chaos». The first large anaerobic fermentation plant was built in 1859 in a leper colony in Bombay [1]. In 1884, Louis Pasteur explored the possibility of obtaining biogas from animal waste and proposed it as a fuel for lighting streetlamps [7]. In general, in the past, biogas was widely used as a source of energy in households in Africa and Asia. Despite the rather primitive design, anaerobic boilers have solved the problems of autonomous energy supply of many households in India, Pakistan, Indochina, and others. Later, biogas became a very important part of the energy source for Western Europe and North America [8].

In today's conditions, the development of biogas production can produce a number of positive effects never before seen, some of which are as follows. First, it is important for modern humanity that biogas energy production can reduce greenhouse gas emissions. In particular, the fermentation of manure on a biogas plant significantly reduces greenhouse gas emissions. The benefit of reducing greenhouse gas emissions is enhanced by the processing and use of methane (a strong greenhouse gas) that could otherwise be released into the atmosphere due to the decomposition of organic by-products and wastes.

Paolini et al. [9], for example, noted that the main goal of the biogas industry is to reduce fossil fuel consumption to mitigate the effects of global warming, and, in general, biogas can make a significant contribution to reducing greenhouse gas emissions. An additional advantage of biogas technology is the production of organic fertilizers for agricultural crops through the use of digestate [10]. In general, biogas can be developed on a large-scale die to partnerships between the energy and agricultural sectors. Another argument in favor of biogas is that it is an environmentally friendly renewable energy source. The only time when biogas is depleted is when the production of any type of waste is stopped. It is also a free source of energy [11]. It is produced by anaerobic biodegradation of organic substances. To do this, the waste must enter an environment without oxygen. This can happen naturally or as part of an industrial biogas process [12–14].

The following also apply:

- Biogas can provide systemic benefits of natural gas (storage, flexibility, high temperature heat) without net carbon emissions. Given the development of a carbon-free economy that humanity is committed to, this is one of the crucial arguments in favor of biogas production.
- Biogas provides a sustainable supply of heat and electricity that can be used by people looking for local, decentralized energy sources, and biogas can be a valuable fuel for cooking in developing countries. In many parts of the world, access to electricity is limited, which makes their way of life more complicated; biogas can provide a good alternative because it is economical to install and possible for both small-scale and large-scale production. Biogas can be used in boilers to produce heat [7,12]. In general, as stated on the website of the World Biogas Association, biogas contributes to the UN Sustainable Development Goal 7: Ensure access to affordable, reliable, sustainable and modern energy for all [15].
- Biogas can play an important role in waste management, increasing overall efficiency of resource use. By converting a number of organic wastes into more valuable products, biogas fits well into the concept of a closed-loop economy. Scholars and practitioners often point out that a constant closed-loop economy can develop largely through the utilization of biomass through the processing of organic waste and, thus, the creation of bioenergy [16–18].
- In cases where biogas displaces gas transported or imported over long distances, it also provides energy security benefits.
- Biogas production can help create jobs in rural areas and reduce the amount of time people spend looking for firewood [11,19,20].

The European Biogas Association's focus on biogas production is a stimulus for rural district development [6]. We consider this aspect to be extremely important for the modern world. In particular, the combination of agricultural activities with renewable energy production using biogas provides three additional benefits: it helps farmers effectively manage their waste and residues, reduces emissions from agriculture, and improves soil quality and biodiversity on agricultural land. In such healthy ecosystems, plants absorb carbon dioxide from the atmosphere, acting as carbon sinks; the digestate used as organic fertilizer returns nutrients to the soil; and methane emissions from livestock enter the controlled environment of the biogas plant, not the atmosphere. Cucui et al. [21] also emphasized that biogas production has a favorable economic effect and can solve waste problems.

The close relationship between agricultural and biogas production is evidenced by the practice of growing cover crops, part of an agronomy system in which an additional second crop is grown before or after the main crop is harvested on the same agricultural land. Cover culture prevents soil erosion and compaction, and promotes the biological, chemical, and physical activity of the soil. As a result, soil quality and fertility are improved, and soils become more resistant to floods and droughts. Cover crops are not ordinary winter crops or pastures, but are sown specifically to protect bare soil in winter and spring after harvesting spring crops. In addition to protecting the soil and its nutrients, cover crops can be of economic importance when used for renewable energy production, such as biogas production.

We fully share the view that bioenergy production from agricultural waste streams and cover crops creates additional business models in the agricultural sector, making agriculture more cost-competitive [6].

In general, in today's information environment there are a lot of data on the benefits of biogas. In our opinion, the most complete set of them is set out, summarized, systematized, and structured in the 2019 report of the World Biogas Association «Global Potential of Biogas». In particular, in this document, the benefits of biogas production are presented in the following main areas: (1) production of renewable energy; (2) climate change mitigation; (3) contribution to the development of the closed-loop economy; (4) improving air quality in cities, the state of water bodies, and soils; (5) contribution to food security; (6) improving health and sanitation through improved solid waste management; (7) economic development and job creation. Moreover, the report emphasizes that in addition to contributing to the Sustainable Development Goals, the development of organic waste processing in general and biogas production in particular is characterized by the following benefits: (1) the possibility of using diverse and local raw materials; (2) flexibility of scaling—biofuels do not have a minimum scale of realization, and their maximum scale is limited only by the amount of available raw materials; (3) flexible use of biogas—primarily for the production of heat and electricity, as well as fuel for cars; (4) multiple income streams—the source of income can be any main or by-product obtained after waste recycling [22].

Indeed, different types of raw materials are used for biogas production. Scientists and practitioners mostly distinguish four groups of raw materials: crop residues—such as wheat stalks and soybean straw; manure from animals (including cattle, pigs, poultry, and sheep); organic fraction of solid household waste, including industrial waste and sewage sludge [19].

The group of plant nutrient residues includes residues after harvesting wheat, corn, rice, and other cereals; and sugar beets, sugar cane, soybeans, and other industrial and oily crops. The group of solid household waste includes food, paper, cardboard, and wood that are not used in other ways (for example, for composting or processing), and some industrial waste from the food industry. Importantly, some industries, such as food, beverages, and chemicals, produce wet waste with a high content of organic substances, which is a suitable raw material for anaerobic digestion. In such industries, biogas production can have the side benefit of cleaning up waste as well as providing heat and electricity on site. The raw material for biogas production is sewage sludge: a semi-solid organic substance extracted in the form of waste gases from municipal treatment plants. In today's conditions, the largest volumes of biogas are produced from agricultural crops and animal manure.

The development of biogas production in the modern world is quite rapid. Thus, in 2010–2018, due to new directions in biogas use, the volume of biogas production increased by 90% worldwide and further growth is expected, especially based on the use of agricultural crop waste. At the same time, the development of biogas production in the world is uneven, as it depends primarily on the availability of raw materials and on policies that encourage its production and use. A total 90% of world production is accounted for by Europe, China, and the United States. In June 2018, EU institutions agreed on a new Renewable Energy Directive for the next decade, including a legally binding EU-wide target to bring the share of renewable energy to 32% by 2030. Given the above benefits of biogas, it is no exaggeration to say that the biogas sector will contribute to this goal [23].

Considering biogas as the most viable option for obtaining energy from organic waste, it is worth noting that, in fact, there are different technologies to produce biomass, such as gasification (energy conversion performances during the biomass air gasification process under microwave irradiation), pyrolysis (microwave-assisted co-pyrolysis of brown coal and corn stover for oil production), etc. The methane content is significantly varied by many factors [24,25]. These are also promising areas of biomass production, which are increasingly implemented in practice.

It is also worth noting that the biogas sector in Europe, as noted by Vlatka Petravie-Tominac, Nikola Nastav, Mateja Buljubasic, and Bozidar Santek, is quite diverse [26].

In particular, it is well developed in Germany, Denmark, Austria, and Sweden, followed by the Netherlands, France, Spain, Italy, The UK, and Belgium. It is characteristic of the EU countries that they have structured their financial incentives in favor of different types of raw materials depending on national priorities—that is, whether biogas production is considered primarily as a way of processing waste (for example, in the UK, more than 80% of biogas is obtained from landfills and sewage sludge), as a way of obtaining renewable energy (for example, in Germany, 93% of biogas is obtained from agricultural crops and agricultural waste), or their combinations. In other EU countries, different combinations of raw materials are used, depending on the specific circumstances and the availability and prices of certain types of raw materials. For example, Denmark predicts gas networks will use only biogas in 2034. At the same time, a significant part of biogas is produced from corn, beets, and animal manure [27]. Aware of the potential of biogas, France has already invested in the technologies needed to develop it on its territory. As of 31 December 2021, for example, the country counted 365 installations designed to inject biomethane into natural gas networks, with a capacity of 6.4 terawatt-hours (TWh) per year [28].

The war in Ukraine has given renewable gas a new impetus, with the European Commission proposing to ramp up biomethane production to 35 billion cubic meters (bcm) by 2030, up from 3 bcm in 2020. The advantage of renewable gas is that it can be produced within Europe. In fact, the actions of many modern European institutions are aimed at the following: (1) making European energy and climate policy constant, stable, and inclusive, increasing the competitiveness of renewable energy sources at cost; (2) increasing the production and consumption of constant renewable gas in Europe for all energy sectors, including transport, domestic, and industrial use; (3) ensuring cross-border and intersectoral exchange of renewable gases; (4) achieving recognition of additional domestic socio-economic and environmental benefits of renewable gas production; (5) making anaerobic fermentation an integral part of competitive and constant agriculture and waste management [29–32].

At the same time, the situation with biogas production in different EU countries depends on a number of different factors, such as investment attractiveness (construction of new biogas plants or only modernization of existing ones); guaranteed green tariff price, which is much higher than electricity produced from other sources; national targets; and action plans for renewable energy sources. In addition, with regard to biogas production, each country has its own obstacles that need to be overcome. One such country is Ukraine, a country that is not self-sufficient in terms of energy supply and is characterized by dependence on low-diversified energy imports, which regularly leads to political tensions and is of high socio-economic importance. At the same time, Ukraine is a powerful agrarian state and is a world leader in the production of numerous kinds of agricultural products. Therefore, in today's Ukrainian society, it is widely believed that the production of agricultural biogas is likely to slow down climate change and increase energy self-sufficiency of the state by replacing or supplementing traditional energy sources [33–35].

Evidence for the development of biogas production in Ukraine being given considerable attention is, for example, the functioning of the Bioenergy Association of Ukraine. In his speeches and publications, the Chairman of the Board emphasizes the need to develop biomethane production, which is in line with the idea of a circular economy, as it converts waste streams, such as agricultural by-products or household waste, into energy while providing recycling of nutrients to agricultural land [35]. In turn, analyst Mostova M., analyzing the current state and prospects of biogas in Ukraine, gave impressive figures: in 2012–2019, in biogas capacity in Ukraine was invested about 140 million euros, and the growth rate of biogas capacity in Ukraine in 2019 was almost 3.5 times higher than in 2018 [36]. As a rule, the production of agricultural biogas is analyzed and evaluated from three main points of view: environmental, economic, and social. Meanwhile, in Ukraine, despite the existing understanding of the importance of biogas use and some efforts by government and business, the growth of total renewable energy production, in particular from biomass, is lagging behind the expected growth. This necessitates constant attention and in-depth research by scientists and practitioners.

The purpose of this publication is to try to model the relationship between the output of biogas from different types of organic mass and to specify the factors that affect it.

## 2. Materials and Methods

To achieve the research goal, the following general scientific research methods were used: theoretical—analyses that included qualitative and quantitative approaches, generalizations, and explanations; empirical—such as description and experiment, particularly modeling; special economic and mathematical methods—factor and correlation analysis.

The data collection technique used for this study consisted of a thorough search for information through Internet search engines in various sources that are relevant to the research problem. The main source of digital data used in the study were statistical data published in a practical guide for Ukrainian farmers.

With the help of qualitative analysis, relevant research articles indexed in Scopus, Web of Science, Google Scholar, and other databases; statistical data; publications in mass media; and program documents of specialized institutions regarding the development of biogas production were analyzed. A brief overview of the dynamics of biogas production in the modern world and an overview of the current situation at the national and international levels are presented. The arguments in favor of biogas production at this stage of human development are summarized, and attention is focused on the positive effects of using biomass of agricultural origin.

An econometric model was used to model the relationship between biogas output from different types of agricultural crops and the proportion of dry matter in crops within the scope of this study—a logical (usually mathematical) description of what economic theory considers particularly important in the study of a certain problem. The choice of this tool of scientific research is due to the fact that the econometric model is a function or a system of functions that describes the correlation–regression relationship between economic indicators. At the same time, depending on the causal relationships among them, one or more of these indicators is considered a dependent variable and the others an independent variable.

Correlation analysis was applied, which made it possible to solve two main tasks: (1) to describe the dependence of the result characteristic on the factor characteristic using a mathematical equation; (2) to evaluate the closeness or density of the connection between the resulting characteristic and its influencing factors. The relationship between the yield of biogas from various agricultural crops and the proportion of dry matter in crops, as well as the potentially possible production of methane from various agricultural crops, has been established in particular.

## 3. Results and Discussion

Thus, within the framework of this study, a multiple econometric model of the relationship between biogas yield from different types of organic substance (m$^3$/t)—$y$ and dry organic substance content (kg/t)—$x_1$, and the share of possible methane content in organic substance $x_2$ (m$^3$/t) was built. The initial data were generally accepted indicators of biogas yield and methane content of substrates of plant and animal origin [37,38].

First of all, primary data are calculated, such as the average values of the factor and result characteristic and their dispersion (Tables 1 and 2).

The average values are used (Formula (1)) to generalize the signs of the set of significant signs, to compare these signs in different sets, and to study the patterns and development trends of phenomena. Since the individual values of the average sign for each unit of the population are known, simple arithmetic means are calculated.

$$y = \frac{\sum y}{n} = \frac{7664}{15} = 510.93 \tag{1}$$

$$x = \frac{\sum x_1}{n} = \frac{9984}{15} = 665.6$$

$$x = \frac{\sum x_2}{n} = \frac{4538.34}{15} = 302.56$$

The study also used the common tool of dispersion, which is the average square of the deviations of all values of the variable sign from its arithmetic mean. This deviation characterizes the average fluctuations of the sign of the totality caused by individual characteristics of the totality from the average value of the sign (Formulas (2) and (3)):

$$\delta_y^2 = \frac{\sum (y - y)^2}{n} = \frac{2242936.93}{15} = 149529.13 \tag{2}$$

$$\delta_{x_1}^2 = \frac{\sum (x_1 - \overline{x}_1)^2}{n} = \frac{1613657.6}{15} = 107577.17$$

$$\delta_{x_2}^2 = \frac{\sum (x_2 - \overline{x}_2)^2}{n} = \frac{998865.18}{15} = 66591.01$$

$$\delta_y = \sqrt[2]{\delta_y} = \sqrt{149529.13} = 386.69 \qquad (3)$$

$$\delta_{x_1} = \sqrt[2]{\delta_{x_1}} = \sqrt{107577.17} = 327.99$$

$$\delta_{x_2} = \sqrt[2]{\delta x_2} = \sqrt{66591.01}$$

**Table 1.** Estimated data for modeling the relationship between biogas yield and different types of organic mass, source [37,38]; calculations of the authors.

| No., i/o | Organic Substance | $y$ | $x_1$ | $x_2$ | $(y-\overline{y})^2$ | $(x_1-\overline{x_1})^2$ | $x_1^2$ | $yx_1$ | $(x_2-\overline{x_2})^2$ |
|---|---|---|---|---|---|---|---|---|---|
| 1 | Cattle manure (fresh) | 90 | 25 | 45 | 177,184.9 | 1727.2 | 625.0 | 2250.0 | 66,334.9 |
| 2 | Pig manure (with bedding) | 75 | 22.5 | 45 | 190,037.9 | 1941.3 | 506.3 | 1687.5 | 66,334.9 |
| 3 | Bird litter (dry) | 80 | 40 | 44 | 185,703.5 | 705.4 | 1600.0 | 3200.0 | 66,851.0 |
| 4 | Bird litter (fresh) | 100 | 15 | 65 | 168,866.2 | 2658.4 | 225.0 | 1500.0 | 56,432.7 |
| 5 | Brewing waste (fresh) | 125 | 24 | 74,125 | 148,944.5 | 1811.4 | 576.0 | 3000.0 | 52,180.6 |
| 6 | Grain: Wheat | 600 | 87 | 316.8 | 7932.9 | 417.8 | 7569.0 | 52,200.0 | 202.9 |
| 7 | Oats | 500 | 87 | 270.5 | 119.5 | 417.8 | 7569.0 | 43,500.0 | 1027.6 |
| 8 | Corn | 590 | 87 | 311.5 | 6251.5 | 417.8 | 7569.0 | 51,330.0 | 80.4 |
| 9 | Rye | 595 | 87 | 309.4 | 7067.2 | 417.8 | 7569.0 | 51,765.0 | 46.8 |
| 10 | Silage, grass, geek | 170 | 37.5 | 90.1 | 116,235.5 | 844.5 | 1406.3 | 6375.0 | 45,137.4 |
| 11 | Sugar beet shavings (pulp) | 595 | 91.6 | 301.1 | 7067.2 | 627.0 | 8390.6 | 54,502.0 | 2,2 |
| 12 | Fat and grease food waste | 845 | 100 | 422.5 | 111,600.5 | 1118.2 | 10,000.0 | 84,500.0 | 14,386.7 |
| 13 | Fat | 875 | 95 | 595 | 132,544.5 | 808.8 | 9025.0 | 83,125.0 | 85,523.7 |
| 14 | Oil: Rapeseed | 1198 | 99.9 | 814.6 | 472,060.6 | 1111.6 | 9980.0 | 119,680.2 | 262,230.4 |
| 15 | Flax, soy, sunflower | 1226 | 99.9 | 833.7 | 511,320.3 | 1111.6 | 9980.0 | 122,477.4 | 282,093.1 |
| | Together | 7664 | 9984 | 4538.3 | 2,242,936.9 | 16,136.6 | 82,590.1 | 681,092.1 | 159,250.9 |

**Table 2.** Estimated data for modeling the relationship between biogas yield and different types of organic mass, source [37,38]; calculations of the authors.

| No., i/o | Organic Substance | $x_2^2$ | $yx_2$ | $x_1 x_2$ | $y^2$ | $\hat{y}$ | $(\hat{y}-\overline{y})^2$ |
|---|---|---|---|---|---|---|---|
| 1 | Cattle manure (fresh) | 2025.0 | 4050.0 | 1125.0 | 8100.0 | 82.4 | 183,606.6 |
| 2 | Pig manure (with bedding) | 2025.0 | 3375.0 | 1012.5 | 5625.0 | 74.1 | 190,827.2 |
| 3 | Bird litter (dry) | 1936.0 | 3520.0 | 1760.0 | 6400.0 | 131.4 | 144,059.9 |
| 4 | Bird litter (fresh) | 4225.0 | 6500.0 | 975.0 | 10,000.0 | 71.7 | 193,045.0 |
| 5 | Brewing waste (fresh) | 5494.5 | 9265.6 | 1779.0 | 15,625.0 | 111.9 | 159,250.9 |
| 6 | Grain: Wheat | 100,362.2 | 190,080.0 | 27,561.6 | 360,000.0 | 595.2 | 7098.1 |

**Table 2.** *Cont.*

| No., i/o | Organic Substance | $x_2^2$ | $yx_2$ | $x_1x_2$ | $y^2$ | $\hat{y}$ | $(\hat{y}-\overline{y})^2$ |
|---|---|---|---|---|---|---|---|
| 7 | Oats | 73,170.3 | 135,250.0 | 23,533.5 | 250,000.0 | 543.1 | 1034.1 |
| 8 | Corn | 97,044.7 | 183,796.8 | 27,102.2 | 348,100.0 | 589.2 | 6132.4 |
| 9 | Rye | 95,728.4 | 184,093.0 | 26,917.8 | 354,025.0 | 586.9 | 5764.5 |
| 10 | Silage, grass, geek | 8118.0 | 15,317.0 | 3378.8 | 28,900.0 | 174.9 | 112,915.8 |
| 11 | Sugar beet shavings (pulp) | 90,643.1 | 179,136.7 | 27,578.0 | 354,025.0 | 592.8 | 6708.6 |
| 12 | Fat and grease food waste | 178,506.3 | 357,012.5 | 42,250.0 | 714,025.0 | 757.5 | 60,793.8 |
| 13 | Fat | 354,025.0 | 520,625.0 | 56,525.0 | 765,625.0 | 934.9 | 179,738.2 |
| 14 | Oil: Rapeseed | 663,638.3 | 975,938.7 | 81,382.5 | 1,435,204.0 | 1198.4 | 472,557.1 |
| 15 | Flax, soy, sunflower | 695,022.3 | 1,022,091.7 | 83,284.6 | 1,503,076.0 | 1219.8 | 502,468.1 |
| | Together | 2,371,964.2 | 3,790,052.0 | 406,165.6 | 6,158,730.0 | 7664.0 | 2,226,000.2 |

In turn, within the framework of this study, the preconditions of correlation analysis were determined:

(1) The yield of biogas from different types of organic mass really depends on the content of dry organic substance and the share of possible methane content in organic substance;
(2) Variation should be sufficient.

In order to compare the set with different values of average arithmetic and standard deviation, we determined the coefficient of variation—the ratio of the square deviation to the average value of the variable sign (Formula (4)):

$$V_y \frac{\delta_y}{\overline{y}} \times 100 = \frac{386.69}{510.93} \times 100 = 75.68 \qquad (4)$$

$$V_{x_1} = \frac{\delta_{x_1}}{\overline{x_1}} \times 100 = \frac{327.99}{665.6} \times 100 = 49.28$$

$$V_{x_2} = \frac{\delta_{x_2}}{\overline{x_2}} \times 100 = \frac{258.05}{302.56} \times 100 = 85.29$$

Variations of the effective and the second factor sign are very large, with $V_y$, $V_{x2} > 50\%$; the variation of the first factor sign is large, as $V_{x1}$ is in the range of 21–50%.

The homogeneity of the totality was checked using the Tao-criterion ($\tau$) (Formula (5)):

$$\tau_{y_{max}} = \frac{y_{max} - \overline{y}}{\delta_y} = \frac{1226 - 510.93}{386.69} = 1.85 \qquad (5)$$

$$\tau_{y_{min}} = \frac{|y_{min} - \overline{y}|}{\delta_y} = \frac{|75 - 510.93|}{386.69} = 1.13$$

$$\tau_{x_{1max}} = \frac{x_{1max} - \overline{x_1}}{\delta_{x_1}} = \frac{1000 - 665.6}{327.69} = 1.02$$

$$\tau_{x_{1min}} = \frac{|x_{1min} - \overline{x_1}|}{\delta_{x_1}} = \frac{|150 - 665.6|}{327.99} = 1.57$$

$$\tau_{x_{2max}} = \frac{x_{2max} - \overline{x_2}}{\delta_{x_2}} = \frac{833.68 - 302.56}{258.05} = 2.06$$

$$\tau_{x_{2min}} = \frac{|x_{2min} - \overline{x_2}|}{\delta_{x_2}} = \frac{|44 - 302.56|}{258.05} = 1.00$$

Since all $\tau < 3$, the totality is defined as homogeneous.

The next step is to describe the relationship between the effective sign and the factors with the help of the matrix regression equation (Formula (6)):

$$y = a_0 + a_1 x_1 + a_2 x_2 + u \qquad (6)$$

Values of unknown parameters $a_0$, $a_1$, $a_2$ were determined based on the method of least squares (Formula (7))

$$S = \sum_{i=1}^{n} (y - \bar{y})^2 \rightarrow \min \qquad (7)$$

and based on the nest equation (Formula (8))

$$S = \sum_{i=1}^{n} (y - (a_0 + a_1 x_1 + \ldots + a_n x_n))^2 \rightarrow \min \qquad (8)$$

Thus, based on the method of least squares, the following system of normal equations (Formula (9)) was obtained:

$$\begin{cases} a_0 n + a_1 \sum x_1 + a_2 \sum x_2 = \sum y \\ a_0 \sum x_1 + a_1 \sum x_1^2 + a_2 \sum x_1 x_2 = \sum x_1 y \\ a_0 \sum x_2 + a_1 \sum x_1 x_2 + a_2 \sum x_2^2 = \sum x_2 y \end{cases} \qquad (9)$$

The next step is to substitute the values of the unknowns into the system of equations and obtain its solution (Formula (10)):

$$\begin{cases} 15 a_0 + 9984 a_1 + 4538,34 a_2 = 7664/15 \\ 9984 a_0 + 8259008 a_1 + 4061655,7 a_2 = 6810921/998,4 \\ 4538,34 a_0 + 4061655,7 a_1 + 2371964,15 a_2 = 3790051,98/4538,4 \end{cases} \qquad (10)$$

$$\begin{cases} a_0 + 665,6 a_1 + 302,56 a_2 = 510,93 \\ a_0 + 827,22 a_1 + 406,82 a_2 = 682,18 \\ a_0 + 894,97 a_1 + 522,65 a_2 = 835,12 \end{cases}$$

II—I, III—II →

$$\begin{cases} 162 a_1 + 104 a_2 = 171/162 \\ 67,74 a_1 + 115,83 a_2 = 152,94/67,74 \end{cases}$$

$$\begin{cases} a_1 + 0,65 a_2 = 1,06 \\ a_1 + 1,71 a_2 = 2,26 \end{cases}$$

II—I →

$$1,06 a_1 = \frac{1,2}{1,06}$$

$$a_2 = 1,13$$

Therefore, the values of unknown parameters were obtained (Formula (11)):

$$a_2 = 1.13 \qquad (11)$$
$$a_1 = 1.06 = 0.65 \times 1.13 = 0.33$$
$$a_0 = 835.12 - 894.97 \times 0.33 - 522.65 = -51.63$$

According to Formula (11), $a_0$—free member of the regression, which has no economic interpretation but contains everything that is not taken into account in the created dependence; $a_{1,2}$—regression coefficients that characterize the proportion of the factor's influence on the result.

Thus, we obtained the regression equation of the following formula (Formula (12)):

$$y = -51.63_{a_0} + 0.33_{a_1} + 1.13_{a_2} \qquad (12)$$

The correctness of the regression equation was checked by the following regularity (Formula (13)):

$$u = \sum \bar{y} - \sum y = 0 \quad (13)$$
$$u = 7664 - 7664 = 0$$

Based on the regression equation (Figure 1), we can calculate the coefficient of elasticity $E_i$ (Formula (14)), which shows by what percentage the value of the resultant trait will change when the factor trait changes by 1%.

$$E_i = a_i \times \frac{\overline{x_i}}{\bar{y}} \quad (14)$$

$$E_1 = a_1 \times \frac{\overline{x_1}}{\bar{y}} = 0.33 \times \frac{665.6}{510.93} = 0.43\%$$

$$E_2 = a_2 \times \frac{\overline{x_2}}{\bar{y}} = 1.13 \times \frac{302.56}{510.93} = 0.67\%$$

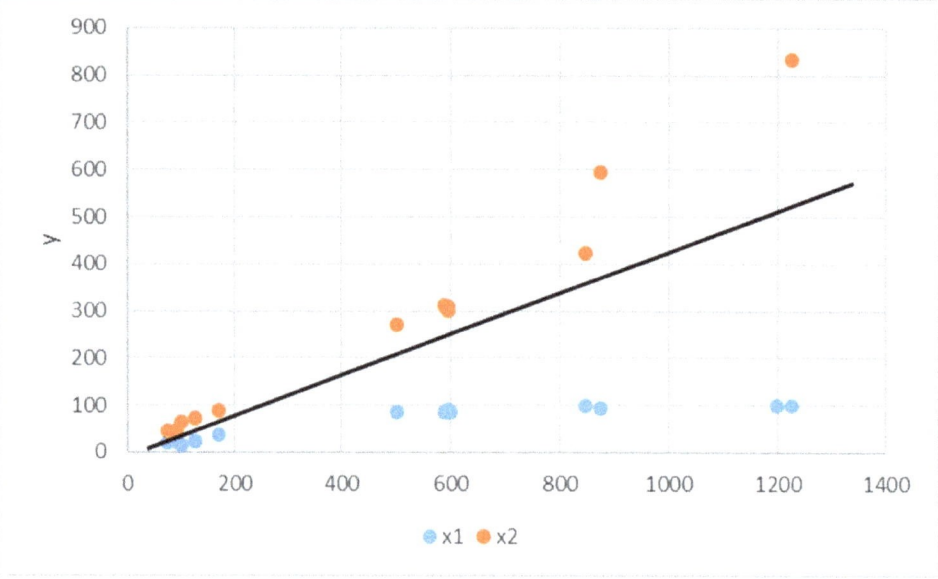

**Figure 1.** Direct regression equitation on the scatter diagram.

Linear correlation analysis is about determining the closeness or density of the relationship between factors and performance. The closeness of the relationship in the correlation analysis is characterized by the correlation coefficient. Accordingly, the density of the relationship between the factors was estimated using simple correlation coefficients (Formula (15)), partial correlation coefficients (Formula (16)), and multiple correlation coefficients (Formula (17)).

Simple correlation coefficients (15):

$$r_{yx_1} = \frac{n \times \sum yx_1 - \sum x_1 \times \sum y}{\sqrt{n \times \sum x_1^2 - (\sum x_1)^2} \times \sqrt{n \times \sum y^2 - (\sum y)^2}} = 0.899 \quad (15)$$

$$r_{yx_2} = \frac{n \times \sum yx_2 - \sum x_2 \times \sum y}{\sqrt{n \times \sum x_2^2 - (\sum x_2)^2} \times \sqrt{n \times \sum y^2 - (\sum y)^2}} = 0.983$$

$$r_{x_1x_2} = \frac{n \times \sum yx_1x_2 - \sum x_1 \times x_2}{\sqrt{n \times \sum x_1^2 - (\sum x_1)^2} \times \sqrt{n \times \sum x_2^2 - (\sum x_2)^2}} = 0.82$$

Partial correlation coefficients (16):

$$r_{yx_1 \times x_2} = \frac{\delta_{yx_1} - \delta_{x_2} \times \delta_{x_1x_2}}{\sqrt{1 - r_{yx_2}^2} \times \sqrt{1 - r_{x_1x_2}^2}} = 0.881 \quad (16)$$

$$r_{yx_2 \times x_1} = \frac{\delta_{yx_2} - \delta_{x_1} \times \delta_{x_1x_2}}{\sqrt{1 - r_{yx_1}^2} \times \sqrt{1 - r_{x_1x_2}^2}} = 0.98$$

Multiple correlation coefficients (17):

$$R_{yx_1x_2} = \sqrt{\frac{r_{yx_1}^2 + r_{yx_2}^2 - 2r_{yx_1} \times r_{yx_2} \times r_{x_1x_2}}{1 - r_{x_1x_2}^2}} \quad (17)$$

To determine the extent to which the constructed econometric model is consistent with the empirical information on the basis of which it was constructed, the coefficient of determination, namely, multiple, was used (Formula (18)):

$$D_{yx_1x_2} = R_{yx_1x_2}^2 \times 100 = 0.966^2 \times 100 = 99.24\% \quad (18)$$

Partial was also used (Formula (19)):

$$d_{yx_1} = a_1 \times r_{yx_1} \times \frac{\delta_{x_1}}{\delta_y} \times 100 = 0.33 \times 0.899 \times \frac{327.99}{386.69} \times 100 = 25.44\% \quad (19)$$

$$d_{yx_2} = a_2 \times r_{yx_2} \times \frac{\delta_{x_2}}{\delta_y} \times 100 = 1.13 \times 0.983 \times \frac{258.05}{386.69} \times 100 = 73.8\%$$

The correctness of the calculation of partial coefficients of determination is checked by regularity (Formula (20)):

$$D_{yx_1x_2} = \sum d_{yx_n} \quad (20)$$

$$99.24\% = 25.44\% + 73.8\%$$

In turn, the multiple factor was checked for materiality. To do this, we formed a null hypothesis $H_0$: $R^2 = 0$ (insignificant) and hypothesis $H_a$: $R^2 \neq 0$ (significant). For checking $H_0$, F-criteria (Fisher's) was used (Formula (21)):

$$F_{R^2} = \frac{\frac{R^2}{p-1}}{\frac{1-R^2}{n-p}} = \frac{\frac{0.992}{3-1}}{\frac{1-0.992}{15-3}} = 788.6 \quad (21)$$

Ftabular = F0.05 =Ftabular < Fcalculated

The significance of partial regression coefficients was assessed using t-criteria (Student's) for the first coefficient (Formula (22)),

$$t_{a_1} = \frac{a_1}{\mu_{a_1}} = \frac{0.33}{0.019} = 17.83 \quad (22)$$

$$\mu_{a_1} = \frac{\delta_{general}}{\delta_{x_1} \times \sqrt{n}} = \frac{\delta_y \times \sqrt{1-R^2}}{\delta_{x_1} \times \sqrt{n}} = 0.019$$

$$t_{a_1} = \frac{a_1}{\mu_{a_1}} = \frac{0.33}{0.019} = 17.83$$

$$\mu_{a_1} = \frac{\delta_{general}}{\delta_{x_1} \times \sqrt{n}} = \frac{\delta_y \times \sqrt{1-R^2}}{\delta_{x_1} \times \sqrt{n}} = 0.019$$

and for the second coefficient (Formula (23)),

$$t_{a_2} = \frac{a_2}{\mu_{a_2}} = \frac{1.13}{0.024} = 47.28 \tag{23}$$

$$\mu_{a_2} = \frac{\delta_{general}}{\delta_{x_2}} = \frac{\delta_y \times \sqrt{1-R^2}}{\delta_{x_2} \times \sqrt{n}} = 0.024$$

To assess the degree of influence of factors on the result, in addition to the correlation coefficient, we calculated it using the multiple correlation index (Formula (24)):

$$\eta_{yx_1x_2} = \sqrt{\frac{\delta^2_{reconstitution}}{\delta^2_y}} = \sqrt{\frac{148400.01}{149529.13}} = 0.966 \tag{24}$$

$$\delta^2_{reconstitution} = \frac{\sum(\acute{y} - \overline{y})^2}{n} = \frac{2226000.2}{15} = 148400.01$$

$$\delta^2_y = \overline{y^2} - \overline{y}^2 = 149529.13$$

## 4. Conclusions

Thus, as a result of building a multiple econometric model of the dependence of biogas yield from different types of agricultural crops on the proportion of dry substance in these crops, and the potential production of methane from them, we obtained the regression equation of the following type:

$$y = -53.61 + 0.3x_1 + 1.13x_2$$

In the regression equation, $a_1 = 0.33$, which means that when the share of dry matter in crops increases by 1 kg/t, the biogas yield will increase by 0.33 m$^2$/t; $a_2 = 1.13$ m$^3$/t means that when the share of methane in biogas increases by 1 m$^3$/t, the biogas yield will increase by 1.13 m$^3$/t.

Since, $a_1$ and $a_2$ are greater than zero, the relationship between the effective sign and the factors influencing it is direct. From this, the coefficient of elasticity $E_1 = 0.43$ means that with an increase in the share of dry substance in agricultural crops of 1%, the yield of biogas will increase by 0.43%; $E_2 = 0.67$ means that with an increase in the share of methane in biogas of 1%, the yield of biogas will increase by 0.67%.

Estimation of the density of the relationship between the factors and the effective sign showed the following results:

(1) Simple correlation coefficients

$R_{yx1} = 0.899$ (strong, positive, direct connection) characterizes the indirect impact on biogas yield not only of the dry substance content of the agricultural crop (kg/t), but also to some extent the impact of the potential share of methane production in the agricultural crop (m$^3$/t);

$R_{yx2} = 0.983$ (strong, positive, direct connection) characterizes the unclean impact on biogas yield not only of the potential share of methane production from agricultural crops (m$^3$/t), but also to some extent the impact of dry substance content in the crop (kg/t);

$R_{x1x2} = 0.820$ (strong, positive, direct relationship) characterizes the closeness of the relationship between the factors influencing the proportion of dry substance in agricultural crops ($x_1$) and potentially possible production of methane from the crop ($x_2$) on the effective basis of biogas output (y).

(2) Partial correlation coefficients

$R_{yx_1} \times x_2 = 0.881$ (strong, positive, direct relationship)—characterizes the value of the net impact of dry substance content in agricultural crops (kg/t) on the yield of biogas from different crops, provided that the factor of the potential share of methane production from the crop (m$^3$/t) is eliminated;

$R_{yx_2} \times x_1 = 0.980$ (strong, positive, direct connection) characterizes the value of the net impact of the potential share of methane production from agricultural crops (m$^3$/t) on the yield of biogas from different crops, provided that the factor of dry substance content in agricultural crops (kg/t) is eliminated.

(3) Multiple correlation coefficients

$Ryx_1x_2 = 0.966$ (strong, positive, direct relationship) characterizes the density (tightness) of the relationship between the biogas output from different agricultural crops and the proportion of dry substance in crops, and the potential for methane production from different agricultural crops.

Checking the closeness of the general relationship (influence) of independent variables on the dependent variable using the coefficient of determination obtained the following results:

(1) Multiple

$Dyx_1x_2 = 99.24\%$ means that the variation in biogas yield from different types of agricultural crops by 99.24% depends on two factors: $x_1$ (share of dry substance in crops) and $x_2$ (potentially possible production of methane from agricultural crops), and 0.76% due to the impact of unaccounted random factors.

(2) Partial coefficients of determination

$d_{yx_1} = 25.44\%$ means that the variation in biogas yield from different types of agricultural crops by 25.44% depends on the factor $x_1$ (share of dry substance in agricultural crops), provided that the factor $x_2$ (potentially possible production of methane from agricultural crops) is eliminated;

$d_{yx_2} = 73.80\%$ means that the variation of biogas yield from different types of agricultural crops by 73.80% depends on factor $x_2$ (potentially possible production of methane from agricultural crops), provided that factor $x_1$ (share of dry substance in agricultural crops) is eliminated.

Fisher's (F) and Student's (t) criteria were used to assess the reliability of correlation characteristics.

As a result of checking the multiple correlation coefficient for materiality using Fisher's F-criteria, its following value was obtained $F_R^2 = 854.3$, which is greater than its tabular value $F_{tabular} = F_{0.05}(2;15) = 3.68$. This means that we reject the null hypothesis and accept the alternative ($H_0: R^2 \neq 0$)—the multiple correlation coefficient is significant.

As a result of estimating the significance of partial regression coefficients using Student's t-criteria, the following results were obtained: $t_{a1} = 17.83$, which is greater than its tabular value $t_{tabular} = t_{0.05}(2;21) = 2.13$. This means that we reject the null hypothesis and accept the alternative ($H_0: t_{a1} \neq 0$)—partial regression coefficient $t_{a1}$ is significant. $t_{a2} = 47.28$, which is greater than its tabular value $t_{tabular} = t_{0.05}(2;21) = 2.13$. This means that we reject the null hypothesis and accept the alternative ($H_0: t_{a2} \neq 0$)—partial regression coefficient $t_{a2}$ is significant.

To estimate the relationship between the effective sign and its factors, a multiple correlation index of 0.966 was calculated. This means that the link between biogas output from different agricultural crops and the proportion of dry substance in crops, and the potential for methane production from different agricultural crops, is strong and functional, as $\eta = 0.966 \to 0$.

Therefore, we consider the results of the study as an argument in favor of further development of biogas production in Ukraine from various types of organic biomass produced by the agricultural sector in order to strengthen energy security. This issue is especially relevant in modern military cataclysms, threats to the energy security of many countries, and the accession of the Ukrainian energy system to the European one. The results of the study are also relevant for other countries where they have issues in the transition of bioenergy.

**Author Contributions:** Conceptualization, N.T. and T.M.; methodology, N.T. and V.N.; validation, N.G., O.T. and O.N.; formal analysis, T.M. and V.N.; writing—original draft preparation, N.T. and N.G.; writing—review and editing, T.M. and V.N.; visualization, O.T. and O.N.; supervision, N.G. and O.T. All authors have read and agreed to the published version of the manuscript.

**Funding:** This research received no external funding.

**Data Availability Statement:** The data presented in this study are available on request from the corresponding author.

**Conflicts of Interest:** The authors declare no conflict of interest.

# References

1. What Is Biogas and Why Is It Important? Howden Group. 2022. Available online: https://www.howden.com/en-gb/articles/renewables/what-is-biogas (accessed on 22 October 2022).
2. Lusiana, L.; Sasongko, N.A.; Kuntjoro, Y.D.; Fakhruddin, M.; Widrian, A.F.; Siregar, A.M.; Vidura, A. Biogas efficiency from cow dung waste in strengthening energy security during the COVID-19 pandemic through a dynamic modeling system. *IOP Conf. Ser. Earth Environ. Sci.* **2021**, *926*, 12085. [CrossRef]
3. Pooja, G.; Goldy, S.; Shivali, S.; Lakhveer, S.; Virendra, K.V. Biogas production from waste: Technical overview, progress, and challenges. *Sustain. Des. Ind. Appl. Mitig. GHG Emiss.* **2020**, *2020*, 89–104. [CrossRef]
4. You Matter. Biogas Definition: What Is Biogas Production? What Are Its Stages? 2020. Available online: https://youmatter.world/en/definition/definitions-biogas-definition-what-is-biogas/ (accessed on 22 October 2022).
5. Energypedia. Advantages and Disadvantages of Biogas. 2021. Available online: https://energypedia.info/wiki/Advantages_and_Disadvantages_of_Biogas (accessed on 22 October 2022).
6. European Biogas Association. About Biogas and Biomethane. 2022. Available online: https://www.europeanbiogas.eu/about-us/vision-mission/ (accessed on 22 October 2022).
7. Korbag, I.; Salma, M.; Hanan, B.; Mousay, A. *Recent Advances of Biogas Production and Future Perspective. Biogas: Recent Advances and Integrated Approaches*; IntechOpen: London, UK, 2020; pp. 1–41. [CrossRef]
8. Damyanova, S.; Beschkov, V. *Biorefinery Concepts, Energy and Products*; Beschkov, V., Ed.; BoD—Books on Demand: Norderstedt, Germany, 2020. [CrossRef]
9. Paolini, V.; Petracchini, F.; Segreto, M.; Tomassetti, L.; Naja, N.; Cecinato, A. Environmental impact of biogas: A short review of current knowledge. *J. Environ. Sci. Health Part A Toxic/Hazard. Subst. Environ. Eng.* **2018**, *53*, 899–906. [CrossRef] [PubMed]
10. Garcia, A. Techno-Economic Feasibility Study of a Small-Scale Biogas Plant for Treating Market Waste in the City of El Alto. Available online: http://www.diva-portal.org/smash/get/diva2:741758/FULLTEXT01.pdf (accessed on 22 October 2022).
11. Shireen, B.; Payal, D. A Review: Advantages and Disadvantages of Biogas. *Int. Res. J. Eng. Technol. (IRJET)* **2017**, *4*, 890–893. Available online: https://www.irjet.net/archives/V4/i10/IRJET-V4I10155.pdf (accessed on 22 October 2022).
12. Raveendran, S.; Parameswaran, B.; Ashok, P.; Snehalata, A.; Yumin, D.; Mukeshm, K.A. Biofuel Production From Biomass: Toward Sustainable Development. *Waste Treat. Process. Energy Gener.* **2019**, *2019*, 79–92. [CrossRef]
13. Mirzoieva, T.; Tkach, N. Economic and production aspects of biomass processing in the energy component. *Bioeco. Agrar. Bus.* **2019**, *10*, 85–96. [CrossRef]
14. Tkach, N.; Mirzoieva, T. Justification of economic performance of processing of grain crops in biogas. *Bioecon. Agrar. Bus.* **2021**, *12*, 5–17. [CrossRef]
15. Biogas' Contribution to the un Sustainable Development Goals, World Biogas Association. 2022. Available online: https://www.worldbiogasassociation.org/goals/energy-security (accessed on 22 October 2022).
16. Abanades, S.; Abbaspour, H.; Ahmadi, A.; Das, B.; Ehyaei, M.; Esmaeilion, F.; El Haj Assad, M.; Hajilounezhad, T.; Jamali, D.H.; Hmida, A.; et al. A critical review of biogas production and usage with legislations framework across the globe. *Int. J. Environ. Sci. Technol.* **2021**, *19*, 3377–3400. [CrossRef]
17. Fagerstrom, A.; Al Seadi, T.; Rasi, S.; Briseid, T. *The Role of Anaerobic Digestion and Biogas in the Circular Economy*; Murphy, J.D., Ed.; IEA Bioenergy Task: Paris, France, 2018; Volume 8, Available online: https://www.ieabioenergy.com/wp-content/uploads/2018/08/anaerobic-digestion_web_END.pdf (accessed on 22 October 2022).
18. Tagne, R.; Dong, X.; Anagho, S.G.; Kaiser, S.; Ulgiati, S. Technologies, challenges and perspectives of biogas production within an agricultural context. The case of China and Africa. *Environ. Dev. Sustain.* **2021**, *23*, 14799–14826. [CrossRef]
19. Outlook for Biogas and Prospects for Organic Growth, World Energy Outlook Special Report. 2020. Available online: https://www.iea.org/reports/outlook-for-biogas-and-biomethane-prospects-for-organic-growth/an-introduction-to-biogas-and-biomethane (accessed on 22 October 2022).

20. Anggraeni, A.; Sahfitri, A.; Hassan, F.; Octavian, A.; Pudjiatmoko, S. Biogas Utilization as Renewable Energy to Achieve National Energy Security (Study of Wonolelo Village, Bantul, Yogyakarta). In Proceedings of the 3rd International Conference of Integrated Intellectual Community, Bantul, Yogyakarta, 27 May 2018. [CrossRef]
21. Sarika, J. Global Potential of Biogasm. World Biogas Associ. 2019. Available online: https://www.worldbiogasassociation.org/wp-content/uploads/2019/07/WBA-globalreport-56ppa4_digital.pdf (accessed on 22 October 2022).
22. Cucui, G.; Ionescu, C.A.; Goldbach, I.R.; Coman, M.D.; Marin, E. Quantifying the Economic Effects of Biogas Installations for Organic Waste from Agro-Industrial Sector. *Sustainability* **2018**, *10*, 2582. [CrossRef]
23. European Biogas Association. Biogas Trends for This Year. 2022. Available online: https://www.europeanbiogas.eu/biogas-trends-for-this-year/ (accessed on 22 October 2022).
24. Ke, C.; Shi, C.; Zhang, Y.; Guang, M.; Li, B. Energy conversion performances during biomass air gasification process under microwave irradiation. *Int. J. Hydrogen. Energy* **2022**, *47*, 31833–31842. [CrossRef]
25. Zhang, Y.; Fan, L.; Liu, S.; Zhou, N.; Ding, K.; Peng, P.; Anderson, E.; Addy, M.; Cheng, Y.; Liu, Y.; et al. Microwave-assisted co-pyrolysis of brown coal and corn stover for oil production. *Bioresour. Technol.* **2018**, *259*, 461–464. [CrossRef]
26. Petravic-Tominac, V.; Nastav, N.; Buljubasic, M.; Santek, B. Current state of biogas production in Croatia. *Energy Sustain. Soc.* **2020**, *10*, 8. [CrossRef]
27. Brooks, C. Denmark Predicts Gas Networks Will Use Only Biogas in 2034. 2022. Available online: https://cleanenergynews.ihsmarkit.com/research-analysis/denmark-predicts-gas-networks-will-use-only-biogas-in-2034.html (accessed on 22 October 2022).
28. Europe Rediscovers Biogas in Search for Energy Independence, EURACTIV France. 2022. Available online: https://www.euractiv.com/section/energy/news/europe-rediscovers-biogas-in-search-for-energy-independence/ (accessed on 22 October 2022).
29. In the Legislative Period 2019–2024 the Eba Will Call for a Common European, European Biogas Association. 2022. Available online: https://www.diva-portal.org/smash/get/diva2:1635090/FULLTEXT01.pdf (accessed on 22 October 2022).
30. Gustafsson, M.; Anderberg, S. Biogas policies and production development in Europe: A comparative analysis of eight countries. *Biofuels* **2022**, *2022*, 931–944. [CrossRef]
31. Perspectives for Biogas in Europe. The Oxford Institute for Energy Studies. 2012. Available online: https://a9w7k6q9.stackpathcdn.com/wpcms/wp-content/uploads/2012/12/NG-70.pdf (accessed on 22 October 2022).
32. Einarsson, R.; Persson, M.; Lantz, M.; Berndes, G.; Cederberg, C.; Toren, J. Estimating the eu Biogas Potential from Manure and Crop Residues—A Spatial Analysis. The Swedish Knowledge Centre for Renewable Transportation Fuels. 2015. Available online: https://f3centre.se/app/uploads/f3_2015-07_einarsson_et_al_final_160107.pdf (accessed on 22 October 2022).
33. Wąs, A.; Sulewski, P.; Gerasymchuk, G.; Stepasyuk, S.; Krupin, K.; Titenko, Z.; Pogodzińska, K. The Potential of Ukrainian Agriculture's Biomass to Generate Renewable Energy in the Context of Climate and Political Challenges—The Case of the Kyiv Region. *Energies* **2022**, *18*, 6547. [CrossRef]
34. Kurbatova, T. Economic benefits from production of biogas based on animal waste within energy co-operatives in Ukraine. *Int. J. Sustain. Energy Plan. Manag.* **2018**, *18*, 69–80. [CrossRef]
35. Geletukha, G. Biomethan—Renewable Gas Which Will Save the Planet. Green Deal. 2021. Available online: https://interfax.com.ua/news/greendeal/742601.html (accessed on 22 October 2022).
36. Mostova, M. The Field of Biogas in Ukraine: Great Prospects and Reality. 2020. Available online: https://energytransition.in.ua/sfera-biohazu-v-ukraini-velyki-perspektyvy-ta-real-nist/ (accessed on 22 October 2022).
37. Sidorchuk, O. Raw Materials for Biogas Production or What Is the Best Way to Harvest Energy. Practical Guide for Farmers "AgroExpert", Section Economics, Series of Articles Biogas—Energy Independence of Ukraine. 2020, Volume 4. Available online: https://agrobiogas.com.ua/raw-materials-for-biogas-production-or-what-best-way-to-harvest-energy/ (accessed on 22 October 2022).
38. Raw Materials for Biogas. AgroBiogas 2020. Available online: https://agrobiogas.com.ua/raw-material-for-biogas/ (accessed on 22 October 2022).

Article

# Activated Carbon from *Stipa tenacissima* for the Adsorption of Atenolol

Nesrine Madani [1], Imane Moulefera [2,*], Souad Boumad [1], Diego Cazorla-Amorós [3], Francisco José Varela Gandía [3], Ouiza Cherifi [4] and Naima Bouchenafa-Saib [1]

[1] Laboratoire de Chimie Physique des Interfaces des Matériaux Appliqués à l'Environnement, Faculté de Technologie, Université Blida 1, B.P. 270 Route de Soumaa, Blida 09000, Algeria
[2] Chemical Engineering Department, Campus Universitario de Espinardo, University of Murcia, 30100 Murcia, Spain
[3] Dept. Química Inorgánica e Instituto Universitario de Materiales, Universidad de Alicante, Ap. 99, 03080 Alicante, Spain
[4] Laboratoire de Chimie du Gaz Naturel, Université des Science et de la Technologie Houarie Boumediene, Bab Ezzouar 16111, Algeria
* Correspondence: imane.moulefera@um.es

**Abstract:** The *Stipa tenacissima S.* is an endemic species of the Western Mediterranean countries, which grows on the semi-arid grounds of North Africa and South Spain. This biomass offers an abundant, renewable, and low-cost precursor for the production of activated carbon (AC). In that context, ACs were prepared by chemical activation of *Stipa tenacissima* leaves (STL) using phosphoric acid ($H_3PO_4$). The effects of activation temperature and impregnation ratio on the textural and chemical surface properties of the prepared activated carbons were investigated. Activation temperatures of 450 and 500 °C turned out to be the most suitable to produce activated carbons with well-developed porous textures. The best results in terms of developed surface area (1503 $m^2/g$) and micropore volume (0.59 $cm^3/g$) were observed for an STLs to phosphoric acid ratio of 1:2 and a carbonization temperature of 450 °C. The adsorption capacity of the optimal activated carbon was found to be 110 mg/g for the atenolol drug. The adsorption equilibrium was well explained by the pseudo-second-order model and Langmuir isotherm. This study showed that the chemical activation method using $H_3PO_4$ as an activating agent was suitable for developing STL-based activated carbon prepared for the removal of atenolol drug in an aqueous solution and compared with commercial activated carbon supplied by Darco.

**Keywords:** activated carbon; *Stipa tenacissima*; chemical activation; adsorption; atenolol

## 1. Introduction

The rise in environmental concerns and pollution issues in recent years has prompted the search for new and sustainable green sources for the production of environmentally friendly materials for environmental applications. The use of biomass as precursors for the production of carbon materials received important attention from many researchers since this is a widely available and abundant source compared to traditional petroleum-based materials, which are polluting, toxic and non-biodegradable [1,2]. Currently, increasing focus is being paid to plant biomass as a raw material, and many industrial companies are following this trend with a major interest in developing economic bio-based products and materials from these renewable materials. Thus, the valorization of biomass into activated carbon is the subject of various works [3–7]. The global activated carbon market is expected to garner 2776 kilotons and 5129 million USD by 2022, registering a compound annual growth rate of 6.83% and 9.32% during the forecast period 2016–2020 [8]. The extensive use of activated carbon is mainly due to its large number of industrial applications, including water and wastewater treatment [9–11], wastewater reclamation [12], gas purification [13–15], or as adsorbents for either $CO_2$ capture or high-pressure $CH_4$

storage [16,17] and also as catalysts [18–20] and catalyst supports [21,22]. Among the large variety of activation processes, chemical activation with phosphoric acid of biomass is one of the most employed methods for the preparation of activated carbon with enhanced physico-chemical properties [23,24]. It presents multiple advantages being the phosphoric acid not toxic compared to other impregnating chemicals [25], the low activating temperature required [25], a high yield obtained [26], and a well-developed mesoporosity [27]. Many factors, during preparation, play an important role in obtaining high-quality activated carbon. The knowledge and control of those variable factors during the activation process is very important in developing the porous texture of the activated carbon that is sought for given applications [9], as this last depends strongly on both, the activation process [16,28] and the nature of the precursor [29–31]. Hence, the influence of the preparation condition parameters such as impregnation ratios and activation temperature was deeply analyzed by many researchers [32,33].

Although there are many applications of activated carbon in different industries, adsorption still remains an effective process that results in extensive use of activated carbons. Consequently, the production of activated carbons with specific pore size distributions from low-cost materials at moderate temperatures is an important challenge on both, economical and energetical aspects [24]. Nowadays, ACs can be produced from a wide range of natural and synthetic substances and lignocellulosic materials being this last one the most used precursors [34–36]. The important content of cellulose and lignin in lignocellulosic materials has promoted them to be the most desired precursors for the preparation of activated carbon, being those two indispensables for getting a high carbon yield [37,38].

Removal of emergent pollutants from wastewater by different methods has been an important challenge for recent society and the subject of several studies over the last years [39–41]. Among them, pharmaceutical products have been widely reported due to their harmful effects on the environment such as paracetamols, clofenac, and some β-blocker species [42–44]. Atenolol is a beta blocker medicament usually used to treat high blood pressure or hypertension, heart rhythm problems, and angina [45]. Around 50% of the dose is not fully metabolized by the human body and is disposed of unchanged through urine [46]. Therefore, it has been extensively detected in concentrations ranging from about 0.78 µg/L to 6.6 µg/L in wastewater and hospital sewage [46,47].

In our previous study [48], the preparation of activated carbons from Stipa tenacissima leaves (STLs), a lignocellulosic plant widely abundant in Southern Algeria, through chemical activation with $H_3PO_4$ has been reported. It has been shown that relatively low temperatures are preferred for the preparation of activated carbons. For this purpose and with the attempt to obtain a well-developed porous texture at low temperatures, this study, on one hand, investigated the preparation of activated carbons from STLs by chemical activation with $H_3PO_4$ at different impregnation ratios (R) and activation temperatures (T). These factors were extensively examined. The study range varied from 400 to 600 °C and from 1 to 3 for both activation temperature and impregnation ratio, respectively. To check further the quality of our obtained ACs, atenolol medicament removal was used as a test to verify the adsorption capacities of three activated carbons, compared with commercial activated carbon from Darco (commercial DARCO G60 derived from lignin delivered by Fluka Chemika (ref. 05100), where the kinetics study was investigated and Freundlich and Langmuir models were reported as well.

## 2. Materials and Methods
### 2.1. Preparation of Activated Carbon

Washed clean STLs (collected from Southern Algeria) were dried in an oven at 110 °C for 24 h, which proved effective to facilitate subsequent crushing and grinding. The precursor was impregnated with an 85% $H_3PO_4$ solution at room temperature and dried for 2 h at 110 °C. The impregnation ratio, R, ($H_3PO_4$/precursor mass) (wt./wt.) was varied from 1 to 3. The samples were activated in a quartz reactor at different temperatures in the range of 400 to 600 °C under nitrogen flow at a rate of 100 mL/min and for 1 h as an activation time. The activated samples were cooled inside the furnace maintaining the $N_2$ flow. After that, the samples were washed with distilled water at 65 °C until neutral pH is achieved before being dried at 110 °C. The obtained activated carbons were named Rx-y, where R from ratio, while x and y correspond to the impregnation ratio and activation temperature, respectively.

## 2.2. Adsorption Equipment and Procedures

Atenolol ($C_{14}H_{22}N_2O_3$) (molecular weight: 266,336 g/mol, melting point: 147 °C, solubility of 13.3 mg/mL (at 25 °C), pKa1 = 9.6; pKa2 = 13.88, polar surface area 84.6 Å2 [49], solutions were prepared with distilled water at different initial concentrations. The equilibrium test was carried out inside glass flasks using 50 mg of dried activated carbon in contact with 100 mL of the atenolol solution at different initial concentrations in an orbital incubator (Gallenkamp, model INR-250) with an equivalent stirring rate of 200 rpm at 25 °C. The kinetic tests of atenolol adsorption were performed using a 100 mL atenolol solution with an initial concentration of 50 mg/L and 50 mg of activated carbon samples under continuous stirring for different time intervals. The concentration of atenolol was analyzed using double beam UV–visible spectrophotometer from Shimadzo (Series UV-1900) at a maximum absorption wavelength of 274 nm.

Adsorption capacity, for each equilibrium concentration, was calculated as is expressed in Equation (1):

$$q_e = \frac{C_0 - C_e}{w} \cdot V \quad (1)$$

where $q_e$ is the equilibrium adsorption capacity (mg/g), $C_0$ and $C_e$ are the initial and equilibrium concentrations, respectively, in mg/L; $V$ is the solution volume (L) and $W$ is the weight of the activated carbon (g).

The equilibrium adsorption data were fitted to the Langmuir, Freundlich, and Temkin adsorption isotherm models (Equations (2), (3) and (4), respectively).

$$q_e = \frac{q_L \cdot K_L \cdot C_e}{1 + K_L \cdot C_e} \quad (2)$$

$$q_e = K_f \cdot (C_e)^{1/n} \quad (3)$$

$$q_e = \frac{RT}{K_1} \cdot \ln(K_2 C_e) \quad (4)$$

where $K_L$ is the equilibrium constant of Langmuir equation (L/mg), $q_L$ is the maximum adsorption capacity (mg/g), $K_f$ is the Freundlich constant associated to the adsorption capacity ((mg/g)(L/mg)$^{1/n}$) and $n$ is the empirical parameter related to the energetic heterogeneity of the adsorption sites, where $K_1$ (J/mol) and $k_2$ (L/mg) is Temkin constant related to the heat of adsorption and isotherm constant, respectively [5].

The adsorption data were fitted to the first-order (Equation (5)), and second-order (Equation (6)) kinetic models for adsorption [48]:

$$Ln(q_e - q_t) = Ln(q_e) - k1t \quad (5)$$

$$\frac{t}{q_t} = \frac{1}{k_2 \cdot q_e^2} + \frac{1}{q_e}t \quad (6)$$

where $k_1$ (L/min) and $k_2$ (g/mg·min) are the kinetic constants for the pseudo-first-order and second-order equation, respectively, and $q_e$ is related to the adsorption capacity at equilibrium (mg/g).

## 2.3. Characterization

Thermal drying method is used for the determination of moisture content of the raw material. The STLs were dried at 110 °C until the consistency of weight was obtained. The moisture percentage in the sample was expressed as the loss in the mass due to drying as a percentage of the total mass of the sample.

Elemental analysis of the precursor and activated carbons was carried out in a CHNS Analyzer. Prior to analysis, the samples were dried overnight at 105 °C and cooled in desiccators. Oxygen content was obtained by the difference between the total percentage (100 wt.%) and the sum of percentages (wt.% dry ash-free) of nitrogen, carbon, hydrogen, and sulfur.

The textural properties of the prepared samples were assessed by nitrogen adsorption–desorption measurements at −196 °C using a Quantachrome Autosorb-6 apparatus. The materials were previously degassed at 250 °C for 4 h. The surface area ($S_{BET}$) was calculated from isotherms using the Brunauer–Emmett–Teller (BET) equation [50]. The volume of liquid nitrogen corresponding to the amount adsorbed at a relative pressure of $P/P_0 = 0.99$ was defined as total pore volume ($V_T$). The micropore volume ($V_{\mu p}$) was determined from Dubinin–Radushkevich equation [51].

## 3. Results and Discussion

### 3.1. Characterization

The results of the proximate analysis are compiled in Table 1. It can be seen that Stipa Tenacissima leaves (STLs) contain 62.81% volatile matter, 24.50% fixed carbon, and 1.19% ash. This composition follows the general trend of a typical biomass composition [34,52–54]. The high volatile matter and low ash content of biomass resources make them good starting materials for preparing activated carbons [55].

**Table 1.** Proximate analysis of Stipa tenacissima leaves (STLs).

| Proximate Analysis | Weight (%) |
|---|---|
| Ash | 1.19 |
| Fixed carbon | 24.50 |
| Volatile matter | 62.81 |
| Moisture | 11.50 |

Table 2 summarizes the elemental composition of the precursor and activated carbons prepared from STLs at different activation temperatures and impregnation ratios. The elemental composition, H/C, and O/C atomic ratios results indicate remarkable chemical changes in the surface after the activation process, while no sulfur (S) traces were detected for all the samples.

**Table 2.** Elemental analysis of the precursor and activated carbons produced at different activation temperatures and impregnation ratios (wt.%).

| Samples | N | C | H | O * | O/C × 10$^2$ | H/C × 10$^2$ |
|---|---|---|---|---|---|---|
| STLs | 1.10 | 47.70 | 6.40 | 44.80 | 93.92 | 13.42 |
| R1-400 | 2.50 | 76.50 | 2.40 | 18.60 | 24.31 | 3.14 |
| R1-450 | 0.50 | 81.60 | 2.10 | 15.80 | 19.36 | 2.57 |
| R1-500 | 0.50 | 82.60 | 1.90 | 15.00 | 18.16 | 2.30 |
| R1-600 | 0.40 | 91.00 | 1.60 | 7.00 | 7.69 | 1.76 |
| R2-400 | 2.30 | 76.20 | 2.00 | 19.50 | 25.59 | 2.62 |
| R2-450 | 0.40 | 78.40 | 1.70 | 19.50 | 24.87 | 2.17 |
| R2-500 | 0.40 | 79.10 | 1.40 | 19.10 | 24.15 | 1.77 |
| R2-600 | 0.30 | 81.60 | 1.40 | 16.70 | 20.47 | 1.72 |
| R3-400 | 1.50 | 70.30 | 1.80 | 26.40 | 37.55 | 2.56 |
| R3-450 | 0.20 | 72.00 | 1.50 | 26.30 | 36.53 | 2.08 |
| R3-500 | 0.20 | 75.20 | 1.30 | 23.30 | 30.98 | 1.73 |
| R3-600 | 0.20 | 78.30 | 1.30 | 20.20 | 25.80 | 1.66 |

(*): by difference.

The results demonstrated that carbon is the major constituent of the obtained ACs confirming the carbonaceous nature of the materials [55]. An increase in the carbon content from 47.74 wt.% for raw STLs to more than 70 wt.% could be observed in all the activated carbons with increasing activation temperature. As for the impregnation ratio of 1, the carbon content in activated carbons increased from 76.46 to 91.04 wt.% with increasing temperatures from 400 to 600 °C, which could be attributed to the increasing release of volatile matter. On another hand, hydrogen and oxygen content highly decreased, respectively, from 2.43 to 1.57 wt.%, and 18.56 to 6.99 wt.%, mainly as a result of the cleavage and breakage of bonds within the ACs structure that occurs during the activation process [56]. Moreover, the progressive decrease in the H/C and O/C atomic fractions (see Figure 1) observed for the obtained activated carbons with the different activation conditions, is indicative of the carbonization and activation processes.

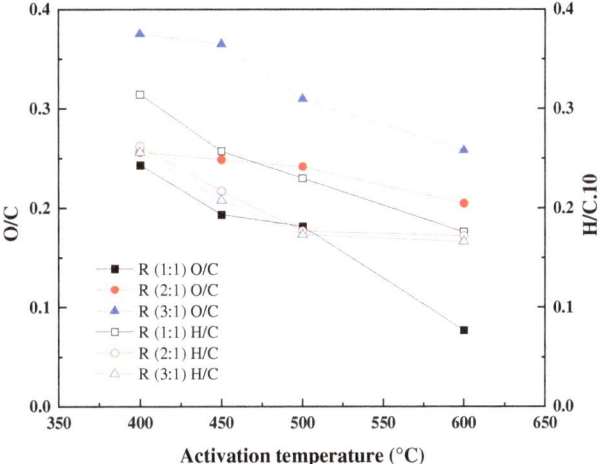

**Figure 1.** Effect of impregnation ratio and activation temperature on H/C and O/C fractions ($H_3PO_4$ concentration: wt. 85%; flow $N_2$ 100 mL/min; activation duration: 1 h).

Indeed, during the activation process, polymeric structures decompose and liberate most of the non-carbon elements, mainly hydrogen, oxygen, and nitrogen in the form of liquid and gases, leaving behind a rigid carbon material with a short-range order [57,58].

Furthermore, as the impregnation ratio increases, carbon, and hydrogen contents decay, whereas the oxygen content increases from 15.53 wt.% for R1-500 to 34.80 wt.% for R3-500. This increase can be due to the progressive incorporation of phosphorus species with increasing the impregnation ratio.

Figure 2a–c, respectively, show the $N_2$ adsorption–desorption isotherms at −196 °C of the prepared activated carbons from STLs with different impregnation ratios and at different activation temperatures. Figure 2a revealed that the isotherms of samples prepared with an impregnation ratio of (1:1) at different activation temperatures are of type I (b) based on IUPAC classification [59], showing a significant increase in the adsorption at low $P/P_0$ values, with barely defined knee, and long plateau which extends to $P/P_0 \approx 1.0$. This is indicative of the presence of large micropores and mesopores. In addition, an absence of hysteresis suggests that the obtained activated carbons contained mostly micropores with only a small contribution of mesopores.

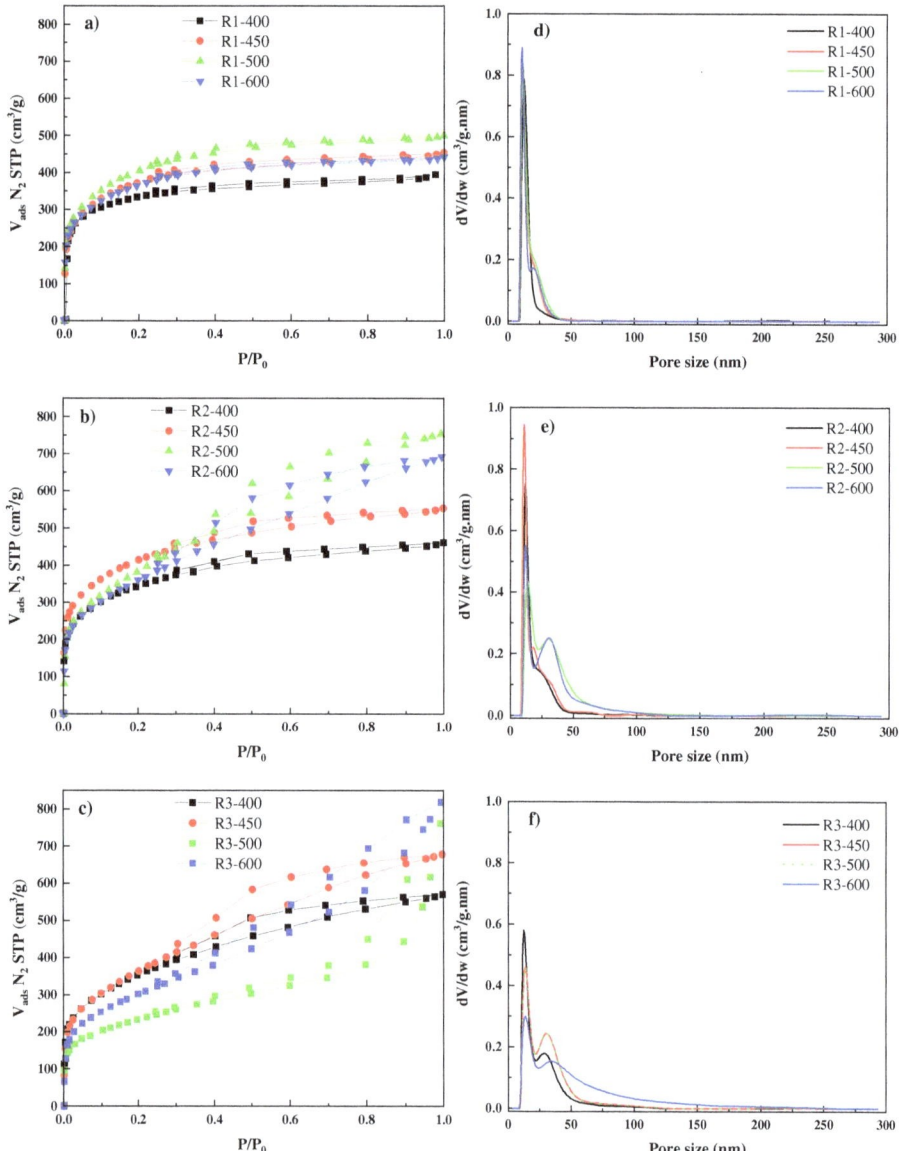

**Figure 2.** Adsorption–desorption isotherms of $N_2$ at −196 °C (**a–c**); and micropore size distribution (**d–f**) of activated carbons from STLs at different impregnation ratios and activation temperatures.

Figure 2b represents the adsorption isotherms of samples prepared with an impregnation ratio of (2:1) at different activation temperatures. The activated carbon obtained at 400 °C provides isotherm type I(b) which is typical of microporous materials where micropore filling may take place by primary filling at very low relative pressure. The activated carbons obtained at higher temperatures, exhibit a combination of type I and type IV(a) isotherms [59]. This indicates the presence of micro and mesoporosity leading to a gradual increase in adsorption after the initial filling of the micropores. The isotherms exhibit type H4 hysteresis, typical for slit-shaped pores.

For the impregnation ratio of (3:1) (Figure 2c) the activated carbons prepared at 400 °C exhibit type I (b) and the activated carbons prepared at 450, 500, and 600 °C a combination of type I and type IV (a) isotherms [59], with the presence of a hysteresis loop type H4. A small hysteresis in the shape was observed in the R3-400 and R3-450 samples. It means that the mesopores are developed during activation with an increasing of impregnation ratio to 3. Additionally, a larger hysteresis loop was observed for R3-500 which suggests a higher contribution of mesopores in their porosity.

The effect of activation temperature and impregnation ratio on the BET surface area, total pores volume, micropores, mesopores volume, and average pore diameter are given in Table 3. The optimum result in terms of surface area (1503 m$^2$/g) was obtained for an impregnation ratio of 2 as can be clearly seen in Table 4. The development of porosity goes through a maximum with the activation temperature, which is typically observed in phosphoric acid activation and is in agreement with our previous results [48]. It is known that phosphoric acid treatment accelerates structural alteration at low temperatures [60]. In fact, it has been reported [61] that at temperatures above 500 °C, the carbon structure shrinks, and the surface area decreases.

**Table 3.** Textural properties of the obtained activated carbons produced at different activation conditions.

| Samples | $S_{BET}$ (m$^2$/g) | $V_{total}$ (cm$^3$/g) | $V_{\mu p}$ (cm$^3$/g) | $V_{meso}$ (cm$^3$/g) | $V_{\mu p}/V_{tot}$ | $D_p$ (nm) |
|---|---|---|---|---|---|---|
| R1-400 | 1204 | 0.61 | 0.53 | 0.07 | 86.88 | 2.03 |
| R1-450 | 1371 | 0.69 | 0.54 | 0.10 | 78.26 | 2.01 |
| R1-500 | 1478 | 0.78 | 0.57 | 0.13 | 73.08 | 2.11 |
| R1-600 | 1340 | 0.69 | 0.52 | 0.11 | 75.36 | 2.06 |
| R2-400 | 1258 | 0.71 | 0.49 | 0.16 | 69.01 | 2.26 |
| R2-450 | 1503 | 0.86 | 0.59 | 0.21 | 68.60 | 2.29 |
| R2-500 | 1387 | 1.17 | 0.53 | 0.53 | 45.29 | 3.37 |
| R2-600 | 1340 | 1.07 | 0.50 | 0.43 | 46.73 | 3.19 |
| R3-400 | 1286 | 0.88 | 0.50 | 0.11 | 56.81 | 2.73 |
| R3-450 | 1317 | 1.05 | 0.47 | 0.45 | 44.76 | 3.18 |
| R3-500 | 1100 | 1.27 | 0.43 | 0.59 | 33.86 | 4.62 |
| R3-600 | 838 | 1.18 | 0.33 | 0.33 | 27.97 | 5.63 |

$S_{BET}$: BET specific surface area; $V_{Total}$: total pore volume; $V_{\mu p}$: micropore volume; $D_p$: average pore diameter.

Girgis et al. have explained that the acid introduced into the material plays a double role [62]: (i) it produces hydrolysis of the lignocellulosic material with subsequent partial extraction of some components, thus weakening the particle which swells, and (ii) the acid occupies a volume which inhibits the contraction of the particle during the heat treatment, thus leaving a porosity when it is extracted by washing after carbonization [62].

Additionally, Jagtoyen et al. have reported that the phosphoric acid combines with organic species forming phosphate and polyphosphate bridges that connect biopolymer fragments and partially hindering the contraction in materials when the temperature increases [63]. Above 450 °C, these bridges become thermally unstable, and their loss produces a contraction in the material, which will result in a decrease in porosity.

From this point of view, keeping the activation temperature at around 500 °C leads to better development of the adsorbent porosity. Several investigators have established that in the case of phosphoric acid activation of lignocellulosic material, temperatures neighboring 500 °C were suitable to obtain optimal properties of the activated carbons.

Impregnation ratio has been identified as one of the most important factors in the chemical activation process. With the increase in ratio from 1 to 3, the surface area and, mainly, total pore volumes also increased. The growth in porosity was attributed to the release of tars from the cross-linked framework generated by the treatment with phosphoric acid [64,65]. In fact, porosity is generated with phosphoric acid remaining in the internal structure of the biopolymer material in the form of phosphate and polyphosphate compounds. As the amount of $H_3PO_4$ used increases, the volume filled by it and various polyphosphates will increase, resulting in larger pore volume and pore size [57].

Figure 2d–f display the micropore size distribution of the different activated carbon obtained from the $N_2$ adsorption at $-196\ °C$. As clearly observed the activated carbon with an impregnation ratio of 1 contain micropores in the range of 12–14 nm (Figure 2d). The increase in the impregnation ratios (Figure 2e,f) results in the appearance of a multimodal pore size distribution.

On the other hand, the experimental data in Table 3 also shows an increase in the percentage of mesoporosity with an increasing impregnation ratio, showing that the development of porosity is also accompanied by a widening of the porosity as the amount of $H_3PO_4$ is increased. These results support those extracted from the above-discussed pore size distribution. When the temperature exceeds 500 °C for the samples prepared with the impregnation ratio of (1:1) and 450 °C for the samples prepared with the impregnation ratio of (2:1) and (3:1), this trend is reversed. This change may be attributed to the increased merging and collapse of micropores which contributes to the reduction of surface area. J. Donald et al. have reported that the phosphate ester cross-links reach their limit of thermal stability at temperatures around 450–500 °C [66]. At higher temperatures, the breakdown of these cross-links would cause contraction and consequent reduction in porosity development.

### 3.2. Adsorption of Atenolol Drug

Three activated carbons were tested to check their removal efficiencies of a pharmaceutical drug "atenolol" from aqueous solution. A commercial activated carbon (CAC) was also used for comparison purposes. The equilibrium and kinetic studies of carbons were investigated. Table 4 contains the textual properties of the activated carbons used for the adsorption of atenolol.

**Table 4.** Textural properties of the tested activated carbons.

| Samples | $S_{BET}$ (m$^2$/g) | $V_{total}$ (cm$^3$/g) | $V_{\mu p}$ (cm$^3$/g) | $V_{meso}$ (cm$^3$/g) |
|---|---|---|---|---|
| R1-500 | 1478 | 0.78 | 0.57 | 0.13 |
| R2-500 | 1387 | 1.17 | 0.53 | 0.53 |
| R3-500 | 1100 | 1.27 | 0.43 | 0.59 |
| CAC | 909 | 0.76 | 0.36 | 0.40 |

#### 3.2.1. Equilibrium Adsorption

The amounts of adsorbed atenolol ($q_e$) against the equilibrium concentration ($C_e$ (mg/L)) at 25 °C are presented in Figure 3. The obtained isotherms are of type L and S for the prepared activated carbons and commercial activated carbon, respectively, according to Giles classification and commonly reported for adsorption in the liquid phase [67]. The L-shape isotherm showed a fairly rapid rise in adsorbed quantity as atenolol concentration increases up to saturation which is characterized by a plateau. This indicates a progressive occupancy of the adsorbent surface as a function of concentration up until the entire surface area is coated with a single layer. Such adsorption behavior could be explained by the high affinity of adsorbent–adsorbate at low and moderate concentrations, which then decreases as concentration increases, since vacant adsorption sites decrease as the adsorbent becomes covered.

The S-shape isotherm showed a small sorption at low concentrations of atenolol in the solution and the sorption increased with the solute concentration. This type of isotherm indicates that at low concentrations the surface has a low affinity for the adsorbate, which increases at higher concentrations because of solute–solute attractive forces.

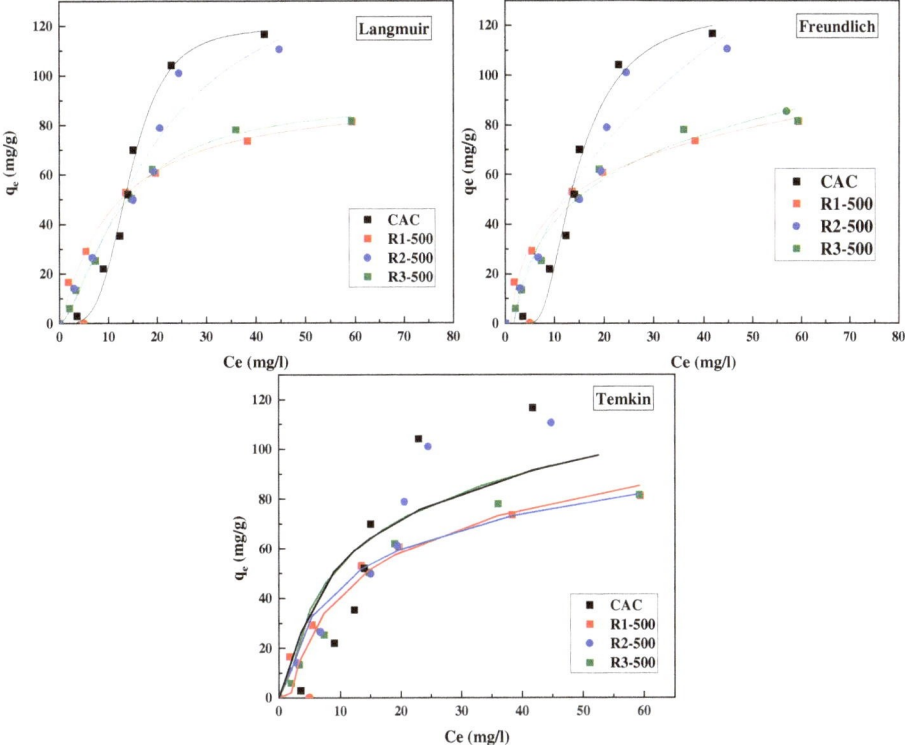

**Figure 3.** Adsorption isotherms of atenolol on the different activated carbons at 25 °C (symbols: experimental values; continuous lines: fitting to Langmuir, Freundlich and Temkin equations).

Furthermore, it can be observed that the adsorption capacity of the R1-500 and R3-500 presented the same amount of about 75 mg/g. However, R2-500 displays a higher adsorption capacity, above 110 mg/g and, with a similar adsorption capacity as for the CAC and this result is much higher compared to the literature [46]. This could mean that a high surface area of the adsorbents may not be the only parameter that determines the removal of higher amounts of the pollutant, but also the distribution of porosity could be a determining factor in this case [68]. This is confirmed by the porosity data (see Table 5), where the R2-500 presents a combination of both a high volume of micropores and an important mesoporous volume.

Three isotherms were used to fit the adsorption experimental results including Langmuir (Equation (2)), Freundlich (Equation (3)), and Temkin (Equation (4)) isotherms, taking into account the effects of equilibrium concentration on adsorption capacity. The three model parameters and correlation coefficients ($R^2$) were listed in Table 5.

It can be observed that in general terms, the highest $R^2$ values were obtained with the Langmuir model for all the tested carbons. The Langmuir adsorption isotherm describes the surface as homogeneous, assuming that there is no lateral interaction between adjacent adsorbed molecules when a single molecule occupies a single surface site. However, considering the shape of the CAC isotherm, this model is not found suitable and could not give the proper information to describe the adsorption process because it does not take into account adsorbate–adsorbate interactions. The maximum monolayer adsorption capacity predicted by the Langmuir model was 98.65, 169.69, and 115.69 mg/g for R1-500, R2-500, and R3-500, respectively.

Table 5. Langmuir, Freundlich, and Temkin adsorption parameters obtained from equilibrium isotherms of atenolol for the activated carbons.

| ACs | R1-500 | R2-500 | R3-500 | CAC |
|---|---|---|---|---|
| **Langmuir isotherm parameters** | | | | |
| $q_L$ (mg/g) | 98.65 | 169.69 | 115.69 | 126.52 |
| $K_L$ (L/mg) | 0.08 | 0.04 | 0.05 | 0.06 |
| $R^2$ | 0.99 | 0.95 | 0.99 | 0.99 |
| **Freundlich isotherm parameters** | | | | |
| $K_F$ (m$^2$/g) | 14.24 | 13.66 | 11.09 | 12.54 |
| 1/n | 0.44 | 0.56 | 0.51 | 0.62 |
| $R^2$ | 0.98 | 0.91 | 0.97 | 0.98 |
| **Temkin isotherm parameters** | | | | |
| $K_1$ (J/mol) | 121.06 | 93.45 | 101.34 | 47.98 |
| $K_2$ (L/mg) | 0.93 | 0.75 | 0.55 | 0.23 |
| $R^2$ | 0.81 | 0.81 | 0.83 | 0.87 |

The extent of the exponent, n, gives information on the favorability of adsorption. As deduced from the results, the values of 1/n were inferior to one (<1), meaning that the adsorption of atenolol was favorable on all samples. Furthermore, $K_F$ is a rough index of the adsorption capacity. A high value of $K_F$ indicates a high adsorption capacity, when the $K_F$ value increases, the adsorption capacity of the adsorbent increases.

3.2.2. Kinetic Study

Figure 4 presents the adsorption kinetics of atenolol on studied carbons. It is clearly observed that the adsorption of AT was faster for samples R1-500 and CAC than that of R2-500 and R3-500 samples and the maximum uptake was reached in approximately 100 min. The quantity adsorbed at equilibrium found in this study (≈65 mg/g) appeared to be better than that reported by N.K Haro et al. (4.0 mg/g) [46].

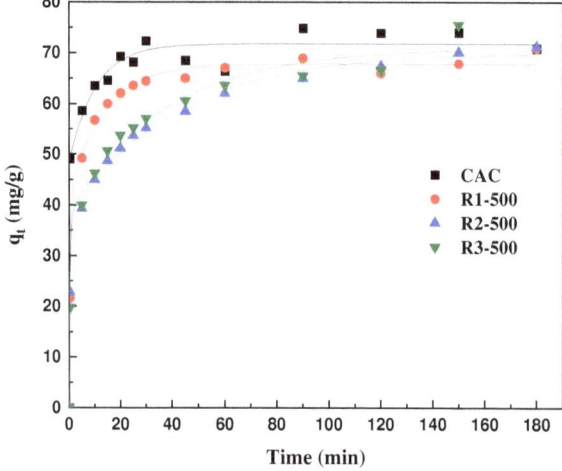

Figure 4. Adsorption kinetics of atenolol on different activated carbons at 25 °C, initial concentration $C_0$ = 50 mg/L; adsorbent concentration = 50 mg/L (experimental data: symbols; and second-order adsorption kinetic equation: continuous line).

Table 6 compiles the fitting parameters for the kinetic studies using (Equations (5) and (6)). The $R^2$ larger than 0.99 as well as the calculated $q_{max}$ values close to the experimental ones indicated that atenolol uptake onto all adsorbents could be satisfactorily described by the pseudo-second-order model. This same tendency was observed in the adsorption of atenolol in a novel β-cyclodextrin adsorbent by Duan et al. [69] and the adsorption of atenolol in biocarbon designed from Melia Azedarach stones by Garcia et al. [70].

**Table 6.** Parameters obtained from kinetics curves of atenolol ($C_0$ = 50 mg/L; adsorbent concentration = 50 mg/L).

| Acs | First-Order Model | | | Experiment | Second-Order Model | | |
|---|---|---|---|---|---|---|---|
| | $q_m$ (mg/g) | $K \cdot 10^2$ (L/mg) | $R^2$ | $q_{exp}$ (mg/g) | $q_m$ (mg/g) | $K \cdot 10^2$ (L/mg) | $R^2$ |
| R1-500 | 18.77 | 1.63 | 0.68 | 70.92 | 69.01 | 6.63 | 0.99 |
| R2-500 | 36.51 | 2.12 | 0.94 | 71.27 | 69.75 | 2.16 | 0.99 |
| R3-500 | 41.61 | 1.78 | 0.84 | 75.53 | 71.53 | 2.07 | 0.99 |
| CAC | 19.74 | 0.20 | 0.68 | 74.06 | 70.45 | 7.62 | 0.99 |

3.2.3. Mechanism of Atenolol Adsorption

The prepared activated carbon is negatively charged, and its surface is rich in oxygenated function. In addition, the pHpzc was found to be about 8. These facts allow the adsorption of atenolol being this last positively charged solution. The species of atenolol would be fixed on the surface of the activated carbon via the interaction H–H and H–O, and the secondary amine group of atenolol as shown in the following proposed mechanism (Figure 5).

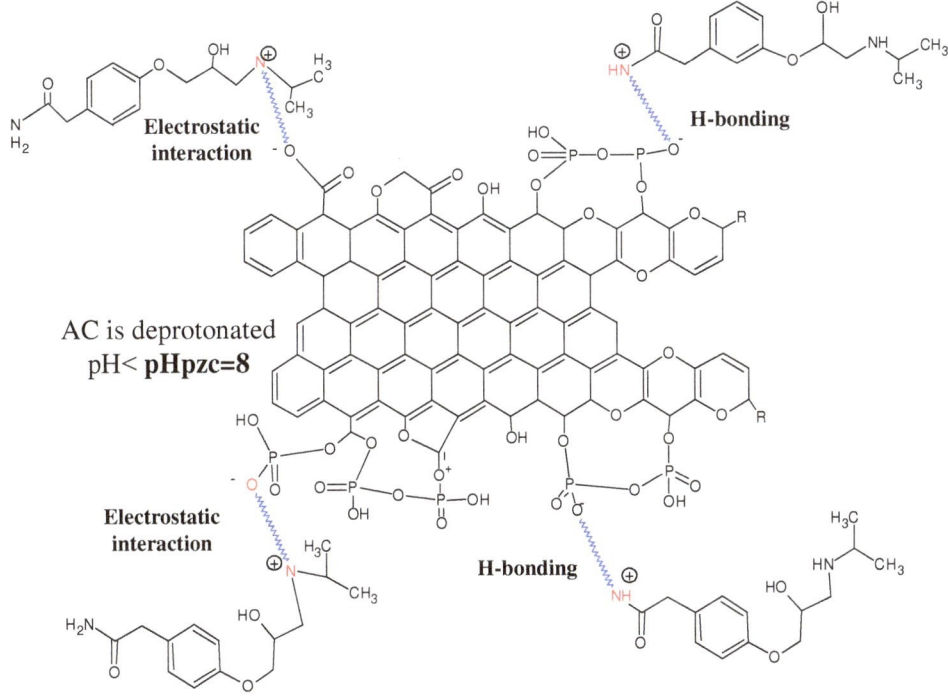

**Figure 5.** Adsorption mechanism of atenolol on activated carbon.

## 4. Conclusions

Activated carbons with a well-developed porosity were prepared from *Stipa tenacisssima* leaves by chemical activation with phosphoric acid at different activation conditions. Generally, increasing the activating agent to STLs ratio from 1 to 3 increased the surface area and, especially, the total pore volume. The obtained results confirm that a 450 °C activation temperature and an impregnation ratio of 2 are suitable for obtaining an activated carbon with a surface area of 1503 m$^2$/g and pore volume of 0.59 cm$^3$/g. The synthesized activated carbons R1-500, R2-500, and R3-500 showed a good adsorption capacity for atenolol removal. The maximum adsorption capacities reached the value of 110 mg/g and showed a similar adsorption capacity as the commercial activated carbon from Darco. The equilibrium and adsorption kinetics results were satisfactorily fitted to Freundlich and Langmuir models and also to the second-order kinetic adsorption equation. Consequently, our findings suggest that a good quality activated carbon could be easily produced by one-step chemical activation with phosphoric acid from cheaper and sustainable raw materials such as *Stipa tenacisssima* leaves, and suitable for the elimination of pharmaceutical drugs and further environmental applications.

**Author Contributions:** N.M.: investigation, formal analysis, writing—original draft preparation; I.M.: methodology, writing—review and editing, formal analysis; S.B.: formal analysis, writing—review and editing; D.C.-A. and F.J.V.G.: formal analysis, reviewing and editing; N.B.-S. and O.C.: conceptualization, project administration, supervision, reviewing and editing. All authors discussed the results and commented on the manuscript. All authors have read and agreed to the published version of the manuscript.

**Funding:** This research received no external funding.

**Institutional Review Board Statement:** Not applicable.

**Informed Consent Statement:** Not applicable.

**Acknowledgments:** The authors thank the Ministry of Higher Education and Scientific Research of Algeria for the internship and the University of Alicante for their cooperation.

**Conflicts of Interest:** The authors declare no conflict of interest.

## References

1. Azwar, E.; Mahari, W.A.W.; Chuah, J.H.; Vo, D.-V.N.; Ma, N.L.; Lam, W.H.; Lam, S.S. Transformation of biomass into carbon nanofiber for supercapacitor application—A review. *Int. J. Hydrogen Energy* **2018**, *43*, 20811–20821. [CrossRef]
2. Moulefera, I.; Trabelsi, M.; Mamun, A.; Sabantina, L. Electrospun Carbon Nanofibers from Biomass and Biomass Blends—Current Trends. *Polymers* **2021**, *13*, 1071. [CrossRef]
3. Nor, N.M.; Lau, L.C.; Lee, K.T.; Mohamed, A.R. Synthesis of activated carbon from lignocellulosic biomass and its applications in air pollution control—A review. *J. Environ. Chem. Eng.* **2013**, *1*, 658–666. [CrossRef]
4. Vargas, A.M.; Cazetta, A.L.; Garcia, C.A.; Moraes, J.C.; Nogami, E.M.; Lenzi, E.; Costa, W.F.; Almeida, V.C. Preparation and characterization of activated carbon from a new raw lignocellulosic material: Flamboyant (Delonix regia) pods. *J. Environ. Manag.* **2011**, *92*, 178–184. [CrossRef] [PubMed]
5. Zhong, Z.-Y.; Yang, Q.; Li, X.-M.; Luo, K.; Liu, Y.; Zeng, G.-M. Preparation of peanut hull-based activated carbon by microwave-induced phosphoric acid activation and its application in Remazol Brilliant Blue R adsorption. *Ind. Crops Prod.* **2012**, *37*, 178–185. [CrossRef]
6. Liou, T.-H. Development of mesoporous structure and high adsorption capacity of biomass-based activated carbon by phosphoric acid and zinc chloride activation. *Chem. Eng. J.* **2010**, *158*, 129–142. [CrossRef]
7. Yahya, M.A.; Al-Qodah, Z.; Zanariah Ngah, C.W. Agricultural bio-waste materials as potential sustainable precursors used for activated carbon production: A review. *Renew. Sustain. Energy Rev.* **2015**, *46*, 218–235. [CrossRef]
8. Allied Market Research. Activated Carbon Market by Product Type (Powdered, Granular), Application (Liquid, Gaseous) and End Use (Water Treatment, Food & Beverage Processing, Air purification)—Global Opportunity Analysis and Industry Forecast, 2014–202. *Allied Mark. Res.* 2022. Available online: https://www.alliedmarketresearch.com/activated-carbon-market (accessed on 4 August 2022).
9. Acharya, J.; Sahu, J.N.; Mohanty, C.; Meikap, B. Removal of lead(II) from wastewater by activated carbon developed from Tamarind wood by zinc chloride activation. *Chem. Eng. J.* **2009**, *149*, 249–262. [CrossRef]
10. Tsai, W.-T.; Lai, C.-W.; Su, T.-Y. Adsorption of bisphenol-A from aqueous solution onto minerals and carbon adsorbents. *J. Hazard. Mater.* **2006**, *134*, 169–175. [CrossRef]

11. García-Mateos, F.; Ruiz-Rosas, R.; Marqués, M.; Cotoruelo, L.; Rodríguez-Mirasol, J.; Cordero, T. Removal of paracetamol on biomass-derived activated carbon: Modeling the fixed bed breakthrough curves using batch adsorption experiments. *Chem. Eng. J.* **2015**, *279*, 18–30. [CrossRef]
12. Kalkan, Ç.; Yapsakli, K.; Mertoglu, B.; Tufan, D.; Saatci, A. Evaluation of Biological Activated Carbon (BAC) process in wastewater treatment secondary effluent for reclamation purposes. *Desalination* **2011**, *265*, 266–273. [CrossRef]
13. Guo, J.; Xu, W.S.; Chen, Y.L.; Lua, A.C. Adsorption of NH3 onto activated carbon prepared from palm shells impregnated with $H_2SO_4$. *J. Colloid Interface Sci.* **2004**, *281*, 285–290. [CrossRef]
14. Sun, Y.; Yang, G.; Zhang, J.; Wang, Y.; Yao, M. Activated Carbon Preparation from Lignin by H3PO4 Activation and Its Application to Gas Separation. *Chem. Eng. Technol.* **2011**, *35*, 309–316. [CrossRef]
15. Dolores, L.-C.; Juan, P.M.-L.; Falco, C.; Titirici, M.-M.; Diego, C.-A. Porous Biomass-Derived Carbons: Activated Carbons. In *Sustainable Carbon Materials from Hydrothermal Processes*; John Wiley & Sons, Ltd.: Hoboken, NJ, USA, 2013; pp. 75–100. [CrossRef]
16. Calvo-Muñoz, E.M.; García-Mateos, F.J.; Rosas, J.; Rodríguez-Mirasol, J.; Cordero, T. Biomass Waste Carbon Materials as adsorbents for CO2 Capture under Post-Combustion Conditions. *Front. Mater.* **2016**, *3*, 1–14. [CrossRef]
17. Ozdemir, S.; Ozdemir, S.; Yetilmezsoy, K. Poultry abattoir sludge as bio-nutrient source for walnut plantation in low-fertility soil. *Environ. Prog. Sustain. Energy* **2019**, *38*, e13066. [CrossRef]
18. García-Mateos, F.J.; Moulefera, I.; Rosas, J.M.; Benyoucef, A.; Rodríguez-Mirasol, J.; Cordero, T. Alcohol Dehydrogenation on Kraft Lignin-Derived Chars with Surface Basicity. *Catalysts* **2017**, *7*, 308. [CrossRef]
19. Matos, I.; Bernardo, M.; Fonseca, I. Porous carbon: A versatile material for catalysis. *Catal. Today* **2017**, *285*, 194–203. [CrossRef]
20. Zhang, S.; Yin, H.; Wang, J.; Zhu, S.; Xiong, Y. Catalytic cracking of biomass tar using Ni nanoparticles embedded carbon nanofiber/porous carbon catalysts. *Energy* **2020**, *216*, 119285. [CrossRef]
21. Szymański, G.; Rychlicki, G. Catalytic conversion of propan-2-ol on carbon catalysts. *Carbon* **1993**, *31*, 247–257. [CrossRef]
22. Arampatzidou, A.C.; Deliyanni, E.A. Comparison of activation media and pyrolysis temperature for activated carbons development by pyrolysis of potato peels for effective adsorption of endocrine disruptor bisphenol-A. *J. Colloid Interface Sci.* **2016**, *466*, 101–112. [CrossRef]
23. Moulefera, I.; García-Mateos, F.J.; Benyoucef, A.; Rosas, J.M.; Rodríguez-Mirasol, J.; Cordero, T. Effect of Co-solution of Carbon Precursor and Activating Agent on the Textural Properties of Highly Porous Activated Carbon Obtained by Chemical Activation of Lignin with $H_3PO_4$. *Front. Mater.* **2020**, *7*, 1–14. [CrossRef]
24. Xu, J.; Chen, L.; Qu, H.; Jiao, Y.; Xie, J.; Xing, G. Preparation and characterization of activated carbon from reedy grass leaves by chemical activation with $H_3PO_4$. *Appl. Surf. Sci.* **2014**, *320*, 674–680. [CrossRef]
25. Kan, Y.; Yue, Q.; Li, D.; Wu, Y.; Gao, B. Preparation and characterization of activated carbons from waste tea by H 3 PO 4 activation in different atmospheres for oxytetracycline removal. *J. Taiwan Inst. Chem. Eng.* **2017**, *71*, 494–500. [CrossRef]
26. Romero, A.; Lillo-Rodenas, M.A.; de Lecea, C.S.-M.; Linares-Solano, A. Hydrothermal and conventional H3PO4 activation of two natural bio-fibers. *Carbon* **2012**, *50*, 3158–3169. [CrossRef]
27. Bouchenafa-Saïb, N.; Mekarzia, A.; Bouzid, B.; Mohammedi, O.; Khelifa, A.; Benrachedi, K.; Belhaneche-Bensemra, N. Removal of malathion from polluted water by adsorption onto chemically activated carbons produced from coffee grounds. *Desalin. Water Treat.* **2013**, *52*, 4920–4927. [CrossRef]
28. Marsh, H.; Rodríguez-Reinoso, F. Preface. In *Activated Carbon*; Marsh, H., Rodríguez-Reinoso, F., Eds.; Elsevier Science Ltd.: Oxford, UK, 2006. [CrossRef]
29. Rodríguez-Reinoso, F.; Molina-Sabio, M. Textural and chemical characterization of microporous carbons. *Adv. Colloid Interface Sci.* **1998**, *76–77*, 271–294. [CrossRef]
30. Nielsen, L.; Biggs, M.J.; Skinner, W.; Bandosz, T.J. The effects of activated carbon surface features on the reactive adsorption of carbamazepine and sulfamethoxazole. *Carbon* **2014**, *80*, 419–432. [CrossRef]
31. Zhang, J.P.; Sun, Y.; Woo, M.W.; Zhang, L.; Xu, K.Z. Preparation of steam activated carbon from black liquor by flue gas precipitation and its performance in hydrogen sulfide removal: Experimental and simulation works. *J. Taiwan Inst. Chem. Eng.* **2016**, *59*, 395–404. [CrossRef]
32. Vernersson, T.; Bonelli, P.R.; Cerrella, E.G.; Cukierman, A.L. *Arundo donax* cane as a precursor for activated carbons preparation by phosphoric acid activation. *Bioresour. Technol.* **2002**, *83*, 95–104. [CrossRef]
33. Jagtoyen, M.; Groppo, J.; Derbyshire, F. Activated carbons from bituminous coals by reaction with H3PO4: The influence of coal cleaning. *Fuel Process. Technol.* **1993**, *34*, 85–96. [CrossRef]
34. Nabais, J.V.; Laginhas, C.; Carrott, M.R.; Carrott, P.; Amorós, J.C.; Gisbert, A.N. Surface and porous characterisation of activated carbons made from a novel biomass precursor, the esparto grass. *Appl. Surf. Sci.* **2013**, *265*, 919–924. [CrossRef]
35. Fierro, V.; Torné-Fernández, V.; Celzard, A.; Montané, D. Influence of the demineralisation on the chemical activation of Kraft lignin with orthophosphoric acid. *J. Hazard. Mater.* **2007**, *149*, 126–133. [CrossRef]
36. Tancredi, N.; Cordero, T.; Mirasol, J.R.; Rodríguez, J.J. Activated carbons from Uruguayan eucalyptus wood. *Fuel* **1996**, *75*, 1701–1706. [CrossRef]
37. Hernández, V. *Lignocellulosic Precursors Used in the Synthesis of Activated Carbon-Characterization Techniques and Applications in the Wastewater Treatment*; InTech: Houston, TX, USA, 2012. [CrossRef]
38. González-García, P. Activated carbon from lignocellulosics precursors: A review of the synthesis methods, characterization techniques and applications, Renew. Sustain. *Energy Rev.* **2018**, *82*, 1393–1414. [CrossRef]

39. Lin, Z.; Wang, D.; Zhang, H.; Li, L.; Huang, Z.; Shen, J.; Lin, Y. Extraction and determination of malachite green from aquatic products based on molecularly imprinted polymers. *Sep. Sci. Technol.* **2016**, *51*, 1684–1689. [CrossRef]
40. Das, A.K.; Saha, S.; Pal, A.; Maji, S.K. Surfactant-modified alumina: An efficient adsorbent for malachite green removal from water environment. *J. Environ. Sci. Health Part A* **2009**, *44*, 896–905. [CrossRef]
41. Farooqi, Z.H.; Sultana, H.; Begum, R.; Usman, M.; Ajmal, M.; Nisar, J.; Irfan, A.; Azam, M. Catalytic degradation of malachite green using a crosslinked colloidal polymeric system loaded with silver nanoparticles. *Int. J. Environ. Anal. Chem.* **2020**. [CrossRef]
42. Zaccariello, G.; Moretti, E.; Storaro, L.; Riello, P.; Canton, P.; Gombac, V.; Montini, T.; Rodríguez-Castellón, E.; Benedetti, A. TiO$_2$–mesoporous silica nanocomposites: Cooperative effect in the photocatalytic degradation of dyes and drugs. *RSC Adv.* **2014**, *4*, 37826–37837. [CrossRef]
43. Al Qarni, H.; Collier, P.; O'Keeffe, J.; Akunna, J. Investigating the removal of some pharmaceutical compounds in hospital wastewater treatment plants operating in Saudi Arabia. *Environ. Sci. Pollut. Res.* **2016**, *23*, 13003–13014. [CrossRef]
44. Rezaei, R.; Aghapour, A.A.; Khorsandi, H. Investigating the biological degradation of the drug β-blocker atenolol from wastewater using the SBR. *Biogeochemistry* **2022**, *33*, 267–281. [CrossRef]
45. Marques, S.C.; Mestre, A.S.; Machuqueiro, M.; Gotvajn, A.; Marinšek, M.; Carvalho, A.P. Apple tree branches derived activated carbons for the removal of β-blocker atenolol. *Chem. Eng. J.* **2018**, *345*, 669–678. [CrossRef]
46. Haro, N.K.; Del Vecchio, P.; Marcilio, N.R.; Feris, L. Removal of atenolol by adsorption—Study of kinetics and equilibrium. *J. Clean. Prod.* **2017**, *154*, 214–219. [CrossRef]
47. Ahmad, A.A.; Din, A.T.M.; Yahaya, N.K.E.; Karim, J.; Ahmad, M.A. Atenolol sequestration using activated carbon derived from gasified Glyricidia sepium. *Arab. J. Chem.* **2020**, *13*, 7544–7557. [CrossRef]
48. Madani, N.; Bouchenafa-Saib, N.; Mohammedi, O.; Varela-Gandía, F.; Cazorla-Amorós, D.; Hamada, B.; Cherifi, O. Removal of heavy metal ions by adsorption onto activated carbon prepared from Stipa tenacissima leaves. *Desalin. Water Treat.* **2017**, *64*, 179–188. [CrossRef]
49. Miroslav, F.; Radka, K.; Oksana, G.; Zuzana, S.; Aleš, K.; Antonín, N.; Martin, K.; Roman, G. Sorption of atenolol, sulfamethoxazole and carbamazepine onto soil aggregates from the illuvial horizon of the Haplic Luvisol on loess. *Soil Water Res.* **2018**, *13*, 177–183. [CrossRef]
50. Brunauer, S.; Emmett, P.H.; Teller, E. Adsorption of Gases in Multimolecular Layers. *J. Am. Chem. Soc.* **1938**, *60*, 309–319. [CrossRef]
51. Dubinin, M.M.; Zaverina, E.D.; Timofeyev, D.P. Sorption and structure of active carbons. I. Adsorption of organic vapors. *Zhurnal Fiz. Khimii* **1947**, *21*, 1352–1362.
52. Sayğılı, H.; Güzel, F.; Önal, Y. Conversion of grape industrial processing waste to activated carbon sorbent and its performance in cationic and anionic dyes adsorption. *J. Clean. Prod.* **2015**, *93*, 84–93. [CrossRef]
53. Demiral, I.; Şamdan, C.A. Preparation and Characterisation of Activated Carbon from Pumpkin Seed Shell Using H$_3$PO$_4$. *Anadolu Univ. J. Sci. Technol. Appl. Sci. Eng.* **2016**, *17*, 125–138. [CrossRef]
54. Tadda, M.A.; Ahsan, A.; Shitu, A.; ElSergany, M.; Arunkumar, T.; Jose, B.; Daud, N.N. A review on activated carbon: Process, application and prospects. *J. Adv. Civ. Eng. Pract. Res.* **2016**, *2*, 7–13.
55. Das, D.; Samal, D.P.; Bc, M. Preparation of Activated Carbon from Green Coconut Shell and its Characterization. *J. Biosens. Bioelectron.* **2015**, *6*, 100248. [CrossRef]
56. Durán-Valle, C.J. Techniques Employed in the Physicochemical Characterization of Activated Carbons. In *Lignocellulosic Precursors Used in the Synthesis of Activated Carbon-Characterization Techniques and Applications in the Wastewater Treatment*; Montoya, V.H., Petriciolet, A.B., Eds.; IntechOpen: London, UK, 2012; pp. 38–58. Available online: https://books.google.dz/books?id=DnmszQEACAAJ (accessed on 4 August 2022).
57. Yakout, S.; El-Deen, G.S. Characterization of activated carbon prepared by phosphoric acid activation of olive stones. *Arab. J. Chem.* **2016**, *9*, S1155–S1162. [CrossRef]
58. Angin, D. Production and characterization of activated carbon from sour cherry stones by zinc chloride. *Fuel* **2014**, *115*, 804–811. [CrossRef]
59. Thommes, M.; Kaneko, K.; Neimark, A.V.; Olivier, J.P.; Rodriguez-Reinoso, F.; Rouquerol, J.; Sing, K.S.W. Physisorption of gases, with special reference to the evaluation of surface area and pore size distribution (IUPAC Technical Report). *Pure Appl. Chem.* **2015**, *87*, 1051–1069. [CrossRef]
60. Jagtoyen, M.; Thwaites, M.; Stencel, J.; McEnaney, B.; Derbyshire, F. Adsorbent carbon synthesis from coals by phosphoric acid activation. *Carbon* **1992**, *30*, 1089–1096. [CrossRef]
61. Kriaa, A.; Hamdi, N.; Srasra, E. Removal of Cu (II) from water pollutant with Tunisian activated lignin prepared by phosphoric acid activation. *Desalination* **2010**, *250*, 179–187. [CrossRef]
62. Girgis, B.S.; Yunis, S.S.; Soliman, A.M. Characteristics of activated carbon from peanut hulls in relation to conditions of preparation. *Mater. Lett.* **2002**, *57*, 164–172. [CrossRef]
63. Jagtoyen, M.; Derbyshire, F. Activated carbons from yellow poplar and white oak by H3PO4 activation. *Carbon* **1998**, *36*, 1085–1097. [CrossRef]
64. Prahas, D.; Kartika, Y.; Indraswati, N.; Ismadji, S. Activated carbon from jackfruit peel waste by H3PO4 chemical activation: Pore structure and surface chemistry characterization. *Chem. Eng. J.* **2008**, *140*, 32–42. [CrossRef]

65. Jun, T.Y.; Latip, N.H.A.; Abdullah, A.M.; Latif, P.A. Effect of Activation Temperature and Heating Duration on Physical Characteristics of Activated Carbon Prepared from Agriculture Waste. *Environ. Asia* **2010**, *3*, 143–148.
66. Donald, J.; Ohtsuka, Y.; Xu, C. Effects of activation agents and intrinsic minerals on pore development in activated carbons derived from a Canadian peat. *Mater. Lett.* **2011**, *65*, 744–747. [CrossRef]
67. Giles, C.H.; Smith, D.; Huitson, A. A general treatment and classification of the solute adsorption isotherm. I. Theoretical. *J. Colloid Interface Sci.* **1974**, *47*, 755–765. [CrossRef]
68. Qu, W.; Yuan, T.; Yin, G.; Xu, S.; Zhang, Q.; Su, H. Effect of properties of activated carbon on malachite green adsorption. *Fuel* **2019**, *249*, 45–53. [CrossRef]
69. Duan, C.; Wang, J.; Liu, Q.; Zhou, Y.; Zhou, Y. Efficient removal of Salbutamol and Atenolol by an electronegative silanized β-cyclodextrin adsorbent. *Sep. Purif. Technol.* **2021**, *282*, 120013. [CrossRef]
70. García-Rosero, H.; Romero-Cano, L.A.; Aguilar-Aguilar, A.; Bailón-García, E.; Carvalho, A.P.; Pérez-Cadenas, A.F.; Carrasco-Marín, F. Adsorption and thermal degradation of Atenolol using carbon materials: Towards an advanced and sustainable drinking water treatment. *J. Water Process. Eng.* **2022**, *49*, 102978. [CrossRef]

*Article*

# Study of Hybrid Modification with Humic Acids of Environmentally Safe Biodegradable Hydrogel Films Based on Hydroxypropyl Methylcellulose

Denis Miroshnichenko [1,*], Katerina Lebedeva [2], Anna Cherkashina [2], Vladimir Lebedev [2], Oleksandr Tsereniuk [3] and Natalia Krygina [3]

1. The Department of Oil, Gas and Solid Fuel Refining Technologies, National Technical University «Kharkiv Polytechnic Institute», 61002 Kharkiv, Ukraine
2. The Department of Plastics and Biologically Active Polymers Technology, National Technical University «Kharkiv Polytechnic Institute», 61002 Kharkiv, Ukraine
3. Pig Breeding Institute and Agro-Industrial Production, National Academy of Agricultural Sciences of Ukraine, 36013 Poltava, Ukraine
* Correspondence: dvmir79@gmail.com

**Abstract:** The possibility of increasing the complexity of the operational properties of environmentally safe biodegradable polymer hydrogel materials based on hydroxypropyl methylcellulose due to modification by humic acids from lignite is considered. As a result of this research, environmentally safe hybrid hydrogel films with antibacterial properties were received. In the framework of physicochemical studies, it was determined by IR spectroscopy that hydroxypropyl methylcellulose modified with humic acids hybridmaterials are received by the mechanism of matrix synthesis, which is accompanied by hydroxypropyl methylcellulose crosslinking through multipoint interaction with the carboxyl group of humic acids. Regularities in terms of changes in water absorption, gelation time, and mold emergence time regarding the environmentally safe biodegradable polymer hydrogel materials based on hydroxypropyl methylcellulose depending on the humic acid content were revealed. It was established that the optimal humic acid content in environmentally safe biodegradable hydrogel films with bactericidal properties based on hydroxypropyl methylcellulose is 15% by mass. It was also established that the hybrid modification of hydroxypropyl methylcellulose with humic acids allows them to preserve their biodegradation properties while giving them antibacterial properties. The environmentally safe biodegradable hydrogel films with bactericidal properties based on hydroxypropyl methylcellulose and humic acids are superior in their operational characteristics to known similar biodegradable hydrogel films based on natural biopolymers.

**Keywords:** environmentally safe; biodegradable; hydrogel films; hydroxypropyl methylcellulose; bactericidal properties; humic acids; hybrid; modification

## 1. Introduction

Plastics and synthetic polymers, which are primarily made from petroleum or petroleum derivatives, have become more common over the years. Their attractive properties, such as their durability, lightness, low cost, and plasticity have contributed to the mass production of plastics in various configurations and their wide application in various industries [1,2]. Plastic is both a blessing and a curse because, despite its properties, plastic persists in the environment for a long time and is easily transported into the biosphere. In 2020, end users in Europe, Norway, Switzerland, and Great Britain threw 29.5 million tons of plastic away, with about 23% of this plastic waste being sent to landfills [3].

Plastics are degraded by sunlight, oxygen, heat, mechanical stress, and/or enzymes into smaller particles such as microplastics and nanoplastics through abiotic and/or biotic degradation [4]. This plastic waste is much more toxic compared to macroplastics.

Numerous studies have reported the role of microplastics as carriers of chemical pollutants, including heavy metals, pesticides, persistent organic pollutants, and persistent bioaccumulative and toxic substances, etc. [5]. Plastic in the soil negatively changes its physical, chemical, and biological properties. Microplastics in the soil can change its porosity, bulk density, and water-holding capacity. The accumulation of microplastics in the soil can also change its biological properties, such as its organic carbon and nitrogen cycling, nutrient transport, and microbial activity [6]. That is why the most relevant direction in industrial polymer materials science today is the receiving of various biodegradable polymers, materials, and composites based on them. Such biodegradable polymers can be broken down abiotically and/or biologically into carbon dioxide, methane, water, and biomass. Such transformations are called biodegradability [7]. Today, there are a large number of biodegradable polymers that implement the principle of "zero waste" during their entire life cycle: «production-use-disposal» [8,9]. The use of a large assortment of environmentally safe biodegradable polymer matrices allows for the obtaining of materials with sufficient strength and heat resistance, which can be processed into various products and parts for various industries. Composite biodegradable materials obtained due to the interaction of chemically different components, most often inorganic and organic, which form a spatial crystal structure that differs from the structures of the original reagents but often inherits the properties of the original components, are called hybrids [10]. Receiving such biodegradable polymer materials and composites allows for a synergistic effect of useful properties for chemically different components in the finished hybrid which leads to area of such materials application expansion.

Several works have shown [11,12] the potential of modifying biodegradable polymer hydrogel materials (BPHMs) with coal, carbon, and graphene oxide derivatives due to their resulting fluorescent ability, photostability, biocompatibility, and large surface areas. Such modified BPHMs are used to receive effective transdermal systems in biomedical applications [13–17]. The most effective modern hydropolymeric microneedle patches are made up of hydroxypropylmethylcellulose [18], hyaluronic acid [19], carboxymethyl cellulose [20], polyvinylpyrrolidone [21], and polylactic glycolic acid [22]. In our opinion, there is great potential for BPHM functional modification based on hydroxypropyl methylcellulose with coal derivatives—humic acids, graphite, graphene, and others. In our previous works, environmentally safe hybrid biodegradable polymer materials based on gelatin [23], polyvinyl alcohol [24], and hydroxypropyl methylcellulose [25], which were modified with humic acids from Ukrainian brown coal [26], were designed and researched. In these works, the antibacterial effect of humic acids in the researched polymers was also established. However, the modification mechanism of hydroxypropyl methylcellulose with humic acids and its impact on the operational characteristics of environmentally safe biodegradable materials with antibacterial action based on them was not determined.

That is why this article studies the hybrid modification of environmentally safe biodegradable hydrogel films based on hydroxypropyl methylcellulose with humic acids. The tasks of this research was to:

- research the physicochemical features resulting from the hybrid modification of environmentally safe biodegradable hydrogels based on hydroxypropyl methylcellulose with humic acids;
- detect the effect of modification with humic acids on a set of strength-based and operational properties regarding environmentally safe, biodegradable hybrid hydrogel films based on hydroxypropyl methylcellulose.

## 2. Materials and Methods

### 2.1. Materials and Reagents

The hydroxypropyl methylcellulose brand was Walocel™, produced by Dow Corning (Dow Corning Inc., Midland, MI, USA). Hydroxypropyl methylcellulose is a natural polymer that dissolves easily and quickly in hot or cold water, forming solutions with different viscosity levels.

The sodium alginate was produced by Qingdao Yingfei Chemical Co (Qingdao Yingfei Chemical Co. Ltd., Shandong, Qingdao, China). Sodium alginate is a linear polysaccharide derivative of alginic acid comprised of 1,4-β-d-mannuronic (M) and α-l-guluronic (G) acids. Humic acids obtained during the extraction of lignite with an alkaline solution of sodium pyrophosphate followed by extraction with a 1% solution of sodium hydroxide and precipitation with mineral acid were used as hybrid modifiers. Table 1 shows the characteristics of the humic acids.

**Table 1.** Proximate analysis of lignite *.

| Proximate Analysis, % wt. | | | |
|---|---|---|---|
| $W^a$ (%) | $A^d$ (%) | $S^d_t$ ($S^{daf}_t$) (%) | $V^{daf}$ ($V^d$) (%) |
| 16.8 | 48.7 | 2.08 (2.50) | 56.7 (29.1) |

* $W^a$—moisture in the analytical state; $A^d$—ash content in a dry state; $S^d_t$ ($S^{daf}_t$)—total sulfur content in a dry state (per organic mass); $V^{daf}$ ($V^d$)—yield of volatile substances per organic mass (in a dry state).

### 2.2. Samples Preparation

Environmentally safe, biodegradable hybrid hydrogel materials based on hydroxypropyl methylcellulose (HESBHM based on hydroxypropyl methylcellulose) were received using the watering method with hydroxypropyl methylcellulose solutions at a concentration of 8% wt. by dissolving the polymer in a mass ratio of 8:100 hydroxypropyl methylcellulose: distilled water heated to 90–100 °C. After that, a defined amount of sodium alginate (2.5% wt.) was added to the previously prepared hydroxypropyl methylcellulose (8% wt.) solution and allowed to mix homogeneously on a magnetic stirrer (magnetic stirrer MM-7P). To analyze the properties of the HESBHM based on hydroxypropyl methylcellulose, solutions of hydroxypropyl methylcellulose and sodium alginate were received at different concentrations of humic acids (5, 10, 15% wt.). A total of 20 parallel experiments were carried out for each composition of HESBHM.

### 2.3. Characterization

IR spectra were obtained on an IR spectrophotometer SPECORD 75 UR at 20–25 °C in the frequency range 4000–500 cm$^{-1}$ under the following conditions: slit—3, recording time—13.2 min., time constant—1 s.

Conductometric studies of polyvinyl alcohol solutions were carried out on a combined TDS-meter HM digital COM-100 (HM Digital Inc., Redondo Beach, CA, USA), scale range:

- Specific conductivity: from 0 to 9990 mkS/cm;
- Temperatures: from 0 to 55 °C;
- Error: ±2%.

Microscopic studies were carried out using the electron microscope Digital Microscope HD color CMOS Sensor (Shenzhen Huahai Hong Communication Technology Co. Ltd., Changzhou, Jiangsu, China).

The viscosity was determined according to ISO 2431. The method is based on determining the viscosity of a solution, with the free flow being taken as the time of continuous flow in seconds of a volume of 50 mL of the test material through a calibrated nozzle with a 4 mm diameter and a VZ-246 viscometer at a certain temperature.

The gelation time was determined by the loss of stickiness time [27].

The water absorption of hydrogel film samples in cold water was carried out according to ISO 62:2008.

Antibacterial properties were determined by the inhibition time of the active growth zones of *Aspergillus niger* (*A. niger*) molds on the surface of HESBHM in a nutrient medium using an electronic microscope, the Digital Microscope HD color CMOS Sensor.

The method described in ISO 846:1997 was used to measure the degree of biodegradation.

## 3. Results and Discussion

*3.1. Rheological and Physical Studies of the Mechanism of Hybrid Modification of Hydroxypropyl Methylcellulose Hydrogels with Humic Acids*

The HESBHM conditional viscosity and conductivity dependence of hydroxypropyl methylcellulose on different humic acid contents is shown in Figure 1.

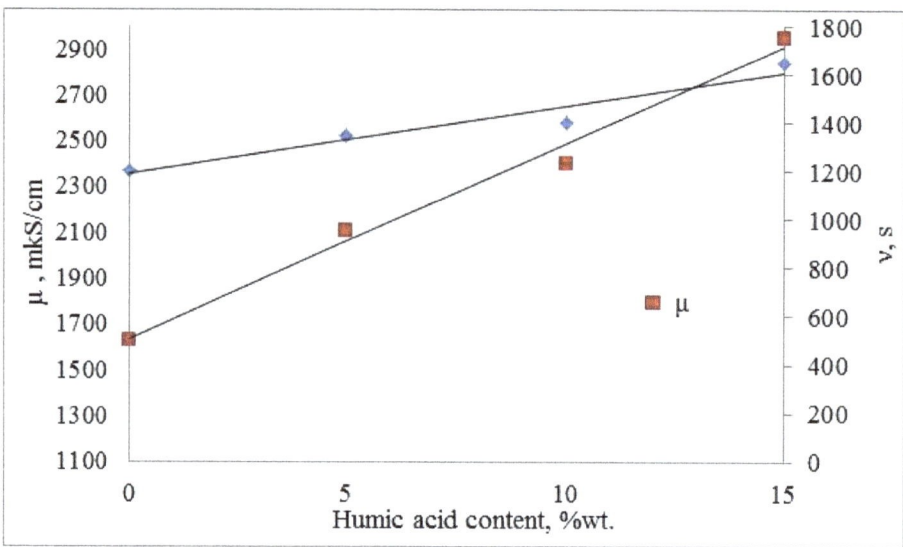

**Figure 1.** The HESBHM conditional viscosity (ν, s), conductivity dependence, and conductivity (μ, mkS/cm) of hydroxypropyl methylcellulose on the different humic acid contents.

From the data in the Figure 1 it can be seen that there is an increasing HESBHM conditional viscosity and specific electrical conductivity with an increase in humic acid content: from 1200 to 1650 s and from 1630 to 2950 mkS/cm, respectively. Such changes indicate the following effects in terms of modification by humic acids on the structure formation processes of HESBHM:

- more density and rigid network in terms of HESBHM, because an increase in viscosity is actually a measure of the density [28] and stiffness of the network formation in water-soluble polymeric hydrogel materials [29,30];
- the formation of a larger number of agglomerates in HESBHM due more intensive hydration process by a high-density and rigid network of HESBHM, because an increase in specific electrical conductivity is actually a measure of the hydration level by the high-density and rigid network in water-soluble polymeric hydrogel materials [31,32].

The formation of a larger number of agglomerates in the HESBHM is clearly visible from the microscopic studies results (Figure 2): unmodified hydroxypropyl cellulose-sodium alginate systems are homogeneous solutions without visible streaks and inhomogeneities both on the surface and in the volume. At the same time, with an increase in humic acid content, the appearance of visible streaks and inhomogeneities both on the surface and in the volume in the HESBHM occurs.

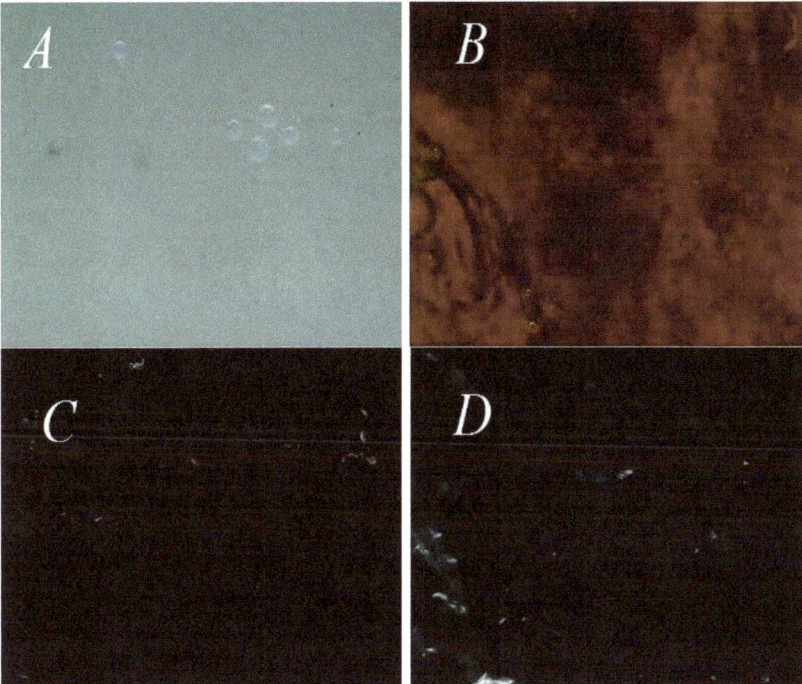

**Figure 2.** Microscopic studies of HESBHM: (**A**) pure hydroxypropyl methylcellulose and sodium alginate hydrogel; (**B**) hydroxypropyl methylcellulose and sodium alginate hydrogel + 5% wt. of humic acid; (**C**) hydroxypropyl methylcellulose and sodium alginate hydrogel + 10% wt. of humic acid; (**D**) hydroxypropyl methylcellulose and sodium alginate hydrogel + 15% wt. of humic acid.

Furthermore, the IR spectra of the original humic acid, original hydroxypropyl methylcellulose, and hydroxypropyl methylcellulose systems (5% by mass) were investigated by IR analysis (Figure 3 and Table 2). It was found that these functional groups determine the humic acid's ability to act as a hybrid modifier in relation to hydroxypropyl methylcellulose. Among the most characteristic humic acid spectral bands are: phenolic –OH hydroxyl groups at 3380–3400 cm$^{-1}$, aliphatic bands C–H at 2920–2940 cm$^{-1}$, symmetric νCOO– carboxyl and νCO (phenolic), and νOH (aliphatic) at 1100 cm$^{-1}$. The IR spectra of the hydroxypropyl methylcellulose–5% wt. of humic acid, characteristic bands from hydroxypropyl methylcellulose and humic acid are clearly observed, for example, a hydroxyl band at 3100–3600 cm$^{-1}$, a methyl band at 2750–2900 cm$^{-1}$, an aromatic C–C band at 1400 and 1600 cm$^{-1}$, a carboxyl band at approximately 1500–1650 cm$^{-1}$, and the C–O band at 1000–1150 cm$^{-1}$ [29]. Compared with the IR spectra of hydroxypropyl methylcellulose and humic acid, there was a significant difference in the IR spectrum of the hydroxypropyl methylcellulose–5 wt%. system of humic acid: a band of carboxyl groups of hydroxypropyl methylcellulose systems–5% by mass of humic acid at 1595 cm$^{-1}$ shifts to wave numbers 1625–1650 cm$^{-1}$. Additionally, it can be seen that there were increases in the formation of hydrogen bonds due to modification, as evidenced by the shift of the hydroxyl band at 3100–3600 cm$^{-1}$ and C–O band at 1000–1150 cm$^{-1}$ to the side by 50–100 cm$^{-1}$.

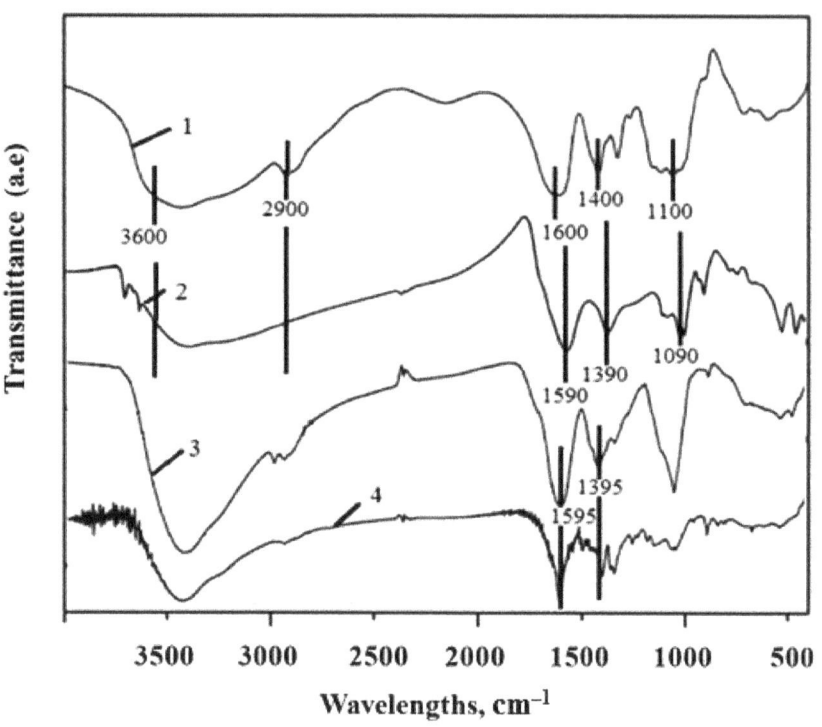

**Figure 3.** IR spectra of humic acids, hydroxypropyl methylcellulose, and hydroxypropyl methylcellulose–humic acids system: 1—hydroxypropyl methylcellulose; 2—humic acids; 3—hydroxypropyl methylcellulose +5% wt. humic acids, 4: hydroxypropyl methylcellulose +10% wt. humic acids.

**Table 2.** IR spectral characteristics of humic acids, hydroxypropyl methylcellulose, and hydroxypropyl methylcellulose–humic acids system.

| Range (cm$^{-1}$) | Functional Groups |
|---|---|
| 3380–3400 | phenolic −OH hydroxyl groups |
| 2920–2940 | aliphatic bands C–H |
| 2750–2900 | −CH$_3$ |
| 1650–1660 | fluctuation $\nu$C=O |
| 1540–1580 | asymmetric $\nu$COO− carboxyl |
| 1400, 1600 | C–C |
| 1380–1400 | symmetric $\nu$COO− carboxyl |
| 1100 | $\nu$CO (phenolic), $\nu$OH (aliphatic) |
| 1040 | $\nu$C–N |
| 1005 | $\nu$CO |
| 910 | out-of-phase $\delta$CH (aromatic) |

Such changes in the IR spectra are evidence that humic acids react with hydroxypropyl methylcellulose through the multipoint interaction of their carboxyl groups with the hydroxyl groups of the polymer, with the formation of such a structure (Figure 4) [18].

Based on the research described above, a general scheme and mechanism for the formation in systems of hydroxypropylcellulose-sodium alginate hybrid modification with humic acids due to the formation of a more rigid network, the enhancement of agglomeration processes, additional supramolecular interactions between functional groups, and an increase in the number of hydrogen bonds are proposed. In fact, the given structure of the hydroxypropyl methylcellulose–humic acid system indicates that it is formed by the mechanism of matrix synthesis within the framework of the hybrid modification of the polymer.

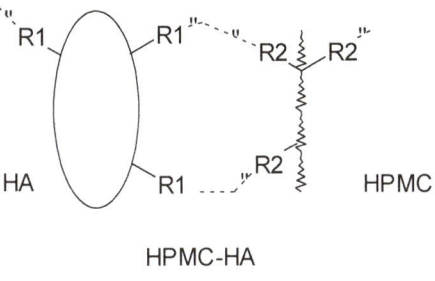

R1 = -OH, -COOH,-PhOH

R2 = -OH

**Figure 4.** The structure of the hydroxypropyl methylcellulose–humic acid system, which is formed by the mechanism of matrix synthesis: HA—humic acids, HPMC—hydroxypropyl methylcellulose.

*3.2. Study of the Effect of Hybrid Modification of Hydroxypropyl Methylcellulose with Humic Acids on a Set of Characteristics of Biodegradable Hydrogel Films*

We have determined that hybrid modification by humic acid changes the most important characteristics of HESBHM: water absorption, gelation time, time of mold appearance, and degree of biodegradability. The graphical dependence of these indicators HESBHM based on hydroxypropyl methylcellulose and the humic acid content of humic acids is shown in Figure 5.

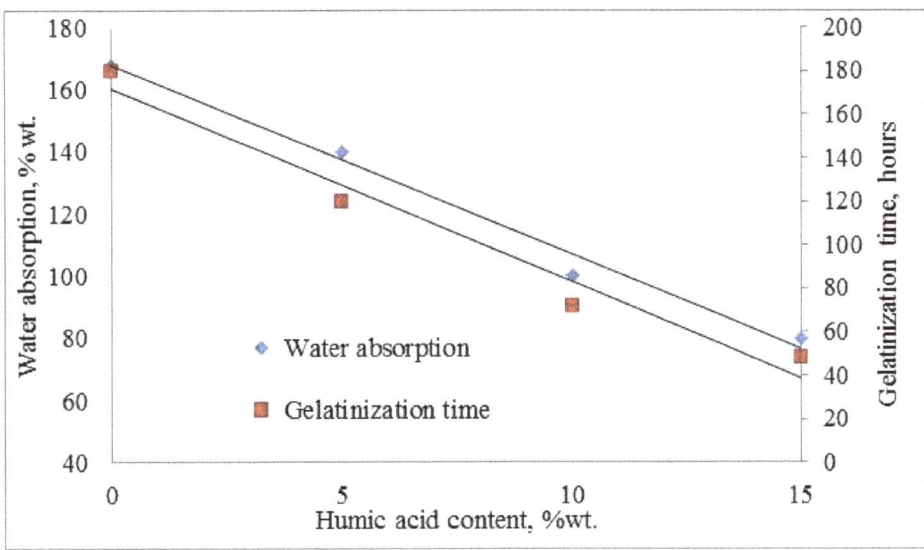

**Figure 5.** Graphical dependence of water absorption and gelation time of HESBHM based on hydroxypropyl methylcellulose and the humic acid content.

From Figure 4, it can be seen that the hybrid modification of hydroxypropyl methylcellulose with humic acids by mechanism matrix synthesis when receiving biodegradable hydrogel films allows for a reduction in their water absorption by reducing the time of gelation.

The mold appearance time and the degree of biodegradation of HESBHM based on hydroxypropyl methylcellulose and the humic acid content are shown in Table 3.

**Table 3.** Characterization of the mold appearance time and the degree of biodegradation of HESBHM based on hydroxypropyl methylcellulose and the humic acid content.

| Sample | Humic Acid Content, % wt. | Degree of Biodegradation, % | Mold Appearance Time (Hours) |
| --- | --- | --- | --- |
| Pure hydroxypropyl methylcellulose and sodium alginate hydrogel | 0 | 99 | 120 |
| Humic acid | 5 | 98 | 350 |
|  | 10 | 97 | does not appear |
|  | 15 | 95 | does not appear |

In general, it was found that the hybrid modification with humic acids according to the mechanism of matrix synthesis of biodegradable hydrogel films based on hydroxypropyl methylcellulose allows for a reduction in their water absorption and gives them antibacterial properties, which is confirmed by the data on the mold appearance time in the films, while preserving their biodegradation properties. It is important to note that the optimal humic substances content for hydroxypropyl cellulose–sodium alginate systems is no more than 15% wt. With this content in terms of humic acids, good antibacterial properties are achieved, due to the complete inhibition of mold formation and sufficiently high levels of water absorption, and the gelation time of hydrogel films are ensured due to an additional supramolecular multipoint interaction occurring between functional groups in addition to hydrogen bonds.

## 4. Conclusions

Environmentally safe hybrid hydrogel films with antibacterial properties were received based on hydroxypropyl methylcellulose modified with humic acids.

Using IR spectroscopy, it was determined that the hydroxypropyl methylcellulose, modified with humic acids, are received by the mechanism of matrix synthesis, which is accompanied by the crosslinking of hydroxypropyl methylcellulose through a multipoint interaction with the carboxyl group in humic acids.

Environmentally safe hybrid hydrogel films based on hydroxypropyl methylcellulose with humic acids (15% wt. content) exhibit antibacterial properties, which are achieved due to the complete inhibition of mold formation, sufficiently high levels of absorption, and the gelation time, which are ensured due to additional supramolecular multipoint interactions occurring between functional groups in addition to hydrogen bonds. The obtained environmentally safe biodegradable hydrogel films with bactericidal properties based on hydroxypropyl methylcellulose and humic substances appear to be promising in terms of their operational characteristics for the preparation of transdermal delivery systems for biologically active substances.

**Author Contributions:** Conceptualization, D.M.; methodology, A.C.; validation, A.C., V.L. and D.M.; formal analysis, K.L.; investigation, K.L.; resources, O.T.; data curation, N.K.; writing—original draft preparation, D.M. and V.L.; writing—review, and editing, D.M. and V.L.; visualization, K.L. All authors have read and agreed to the published version of the manuscript.

**Funding:** This research received no external funding.

**Data Availability Statement:** Not applicable.

**Conflicts of Interest:** The authors declare no conflict of interest.

## References

1. Cabrera, F.C. Eco-friendly polymer composites: A review of suitable methods for waste management. *Polym. Compos.* **2021**, *42*, 2653–2677. [CrossRef]
2. Wang, W.; Ge, J.; Yu, X.; Li, H. Environmental fate and impacts of microplastics in soil ecosystems: Progress and perspective. *Sci. Total Environ.* **2020**, *708*, 134841. [CrossRef] [PubMed]
3. Karamanlioglu, M.; Preziosi, R.; Robson, G.D. Abiotic and biotic environmental degradation on the bioplastic polymer poly(lactic acid): A review. *Polym. Degrad. Stab.* **2017**, *137*, 122–130. [CrossRef]
4. Abbasi, S.; Ammar Haeri, S. Biodegradable materials and their applications in sample preparation techniques–A review. *Microchem. J.* **2021**, *171*, 106831. [CrossRef]
5. Anukiruthika, T.; Sethupathy, P.; Wilson, A.; Kashampur, K.; Moses, J.A.; Anandharamakrishnan, C. Multilayer packaging: Advances in preparation techniques and emerging food applications. *Compr. Rev. Food Sci. Food Saf.* **2020**, *19*, 1156–1186. [CrossRef] [PubMed]
6. Falguera, V.; Quintero, J.P.; Jimenez, A.; Munoz, J.A.; Ibarz, A. Edible films and coatings: Structures, active function and trends in their use. *Trends Food Sci. Technol.* **2011**, *22*, 292–303. [CrossRef]
7. Lebedev, V.; Tykhomyrova, T.; Litvinenko, I.; Avina, S.; Saimbetova, Z. Design and Research of Eco-Friendly Polymer Composites. *Mater. Sci. Forum* **2020**, *1006*, 259–266. [CrossRef]
8. Lebedev, V.; Tykhomyrova, T.; Filenko, O.; Cherkashina, A.; Lytvynenko, O. Sorption Resistance Studying of Environmentally Friendly Polymeric Materials in Different Liquid Mediums. *Mater. Sci. Forum* **2021**, *1038*, 168–174. [CrossRef]
9. Lebedev, V.; Miroshnichenko, D.; Bilets, D.; Mysiak, V. Investigation of Hybrid Modification of Eco-Friendly Polymers by Humic Substances. *Solid State Phenom.* **2022**, *334*, 154. [CrossRef]
10. Kickelbick, G. *Hybrid Materials: Synthesis, Characterization and Applications*, 1st ed.; Wiley-VCH Verlag GmbH & Co.: Weinheim, Germany, 2007.
11. Pyshyev, S.; Demchuk, Y.; Gunka, V.; Sidun, I.; Shved, M.; Bilushchak, H.; Obshta, A. Development of mathematical model and identification of optimal conditions to obtain phenol-cresol-formaldehyde resin. *Chem. Chem. Technol.* **2019**, *13*, 212. [CrossRef]
12. Prysiazhnyi, Y.; Borbeyiyong, G.I.; Pyshyev, S. Preparation and application of coumarone-indene-carbazole resin as a modifier of road petroleum bitumen 1. Influence of carbazole:raw materials ratio. *Chem. Chem. Technol.* **2022**, *16*, 284. [CrossRef]
13. Yang, D.; Hu, Y.; Ma, D.; Ge, J.; Huang, S. Reconfigurable Mechanochromic Patterns into Chameleon-Inspired Photonic Papers. *Research* **2022**, *2022*, 13. [CrossRef] [PubMed]
14. Wang, Y.; Li, J.; Yao, X.; Xie, C.; Chen, Q.; Liu, W.; Gao, Z.; Fu, Y.; Liu, Q.; He, D.; et al. Improved Comprehensive Photovoltaic Performance and Mechanisms by Additive Engineering of Ti3C2Tx MXene into CsPbI2Br. *ACS Appl. Mater. Interfaces* **2022**, *14*, 40930–40938. [CrossRef] [PubMed]
15. Hu, Y.; Yang, D.; Ma, D.; Huang, S. Liquid, Transparent, and Antideformable Thermochromic Photonic Crystals for Displays. *Adv. Opt. Mater.* **2022**, *10*, 2200769. [CrossRef]
16. Hu, Y.; Yang, D.; Ma, D.; Huang, S. Extremely sensitive mechanochromic photonic crystals with broad tuning range of photonic bandgap and fast responsive speed for high-resolution multicolor display applications. *Chem. Eng. J.* **2022**, *429*, 132342. [CrossRef]
17. Song, P.; Wu, Y.; Zhang, X.; Yan, Z.; Wang, M.; Xu, F. Preparation of covalently crosslinked sodium alginate/hydroxypropyl methylcellulose pH-sensitive microspheres for controlled drug release. *BioResources* **2018**, *13*, 8614–8628. [CrossRef]
18. Kim, J.Y.; Han, M.R.; Kim, Y.H.; Shin, S.W.; Nam, S.Y.; Park, J.H. Tip-loaded dissolving microneedles for transdermal delivery of donepezil hydrochloride for treatment of Alzheimer's disease. *Eur. J. Pharm. Biopharm.* **2016**, *105*, 148–155. [CrossRef]
19. Du, H.; Liu, P.; Zhu, J.; Lan, J.; Li, Y.; Zhang, L.; Zhu, J.; Tao, J. Hyaluronic acid-based dissolving microneedle patch loaded with methotrexate for improved treatment of psoriasis. *ACS Appl. Mater. Interfaces* **2019**, *11*, 43588–43598. [CrossRef]
20. Mistilis, M.J.; Bommarius, A.S.; Prausnitz, M.R. Development of a thermostable microneedle patch for influenza vaccination. *J. Pharm. Sci.* **2015**, *104*, 740–749. [CrossRef]
21. Tang, J.; Wang, J.; Huang, K.; Ye, Y.; Su, T.; Qiao, L.; Hensley, M.T.; Caranasos, T.G.; Zhang, J.; Gu, Z.; et al. Cardiac cell-integrated microneedle patch for treating myocardial infarction. *Sci. Adv.* **2018**, *4*, 9365. [CrossRef]
22. Li, Y.; Zhang, H.; Yang, R.; Laffitte, Y.; Schmill, U.; Hu, W.; Kaddoura, M.; Blondeel, E.J.M.; Cui, B. Fabrication of sharp silicon hollow microneedles by deep -reactive ion etching towards minimally invasive diagnostics. *Microsyst. Nanoeng.* **2019**, *5*, 41. [CrossRef] [PubMed]
23. Lebedev, V.; Miroshnichenko, D.; Xiaobin, Z.; Pyshyev, S.; Savchenko, D. Technological Properties of Polymers Obtained from Humic Acids of Ukrainian Lignite. *Pet. Coal* **2021**, *63*, 646–654.
24. Lebedev, V.; Miroshnichenko, D.; Xiaobin, Z.; Pyshyev, S.; Savchenko, D.; Nikolaichuk, Y. Use of Humic Acids from Low-Grade Metamorphism Coal for the Modification of Biofilms Based on Polyvinyl Alcohol. *Pet. Coal* **2021**, *63*, 953–962.
25. Lebedev, V.; Sizhuo, D.; Xiaobin, Z.; Pyshyev, S.; Dmytro, S. Hybrid Modification of Eco-Friendly Biodegradable Polymeric Films by Humic Substances from Low-Grade Metamorphism Coal. *Pet. Coal* **2022**, *64*, 539–546.
26. Miroshnichenko, D.V.; Pyshyev, S.V.; Lebedev, V.V.; Bilets, D.Y. Deposits and quality indicators of brown coal in Ukraine. *Nauk. Visnyk Natsionalnoho Hirnychoho Universytetu* **2022**, *3*, 5–10. [CrossRef]
27. Podgornaya, L.P.; Cherkashyna, G.M.; Lebedev, V.V. *Theory and Methods of Research and Testing of Plastics, Adhesives and Sealants*, 1st ed.; Textbook of NTU "KhPI": Kharkiv, Ukraine, 2012.

28. Feroz, S.; Dias, G. Hydroxypropylmethyl cellulose (HPMC) crosslinked keratin/hydroxyapatite (HA) scaffold fabrication, characterization and in vitro biocompatibility assessment as a bone graft for alveolar bone regeneration. *Heliyon* **2021**, *7*, e08294. [CrossRef]
29. Rizwan, M.; Rubina Gilani, S.; Iqbal Durani, A.; Naseem, S. Materials Diversity of Hydrogel: Synthesis, Polymerization Process and Soil Conditioning Properties in Agricultural Field. *J. Adv. Res.* **2021**, *33*, 15–40. [CrossRef]
30. Cacopardo, L.; Guazzelli, N.; Nossa, R.; Mattei, G.; Ahluwalia, A. Engineering hydrogel viscoelasticity. *J. Mech. Behav. Biomed. Mater.* **2019**, *89*, 162–167. [CrossRef]
31. Kaklamani, G.; Kazaryan, D.; Bowen, J.; Iacovella, F.; Anastasiadis, S.H.; Deligeorgis, G. On the electrical conductivity of alginate hydrogels. *Regen. Biomater.* **2018**, *5*, 293–301. [CrossRef] [PubMed]
32. Konsta, A.; Daoukaki, D.; Pissis, P.; Vartzeli, K. Hydration and conductivity studies of polymer–Water interactions in polyacrylamide hydrogels. *Solid State Ion.* **1999**, *125*, 235–241. [CrossRef]

Article

# Lauric Acid Treatments to Oxidized and Control Biochars and Their Effects on Rubber Composite Tensile Properties

Steven C. Peterson * and A. J. Thomas

USDA, Agricultural Research Service, National Center for Agricultural Utilization Research,
Plant Polymer Research, 1815 N University, Peoria, IL 61604, USA
* Correspondence: steve.peterson@usda.gov

**Abstract:** Biochar is a renewable source of carbon that can partially replace carbon black as filler in rubber composites. Since the carbon content of biochar is less pure than carbon black, improvements and modifications must be made to biochar to make it a viable co-filler. In this work, two methods to change the surface chemistry of biochar were employed: (1) gas treatment at 300 °C with either air or carbon dioxide, and (2) coating with lauric acid. Both methods are amenable to the current rubber processing industry. After biochar was treated with these methods, it was used as co-filler in rubber composite samples. Gas treatment with either air or carbon dioxide was found to increase stiffness in the final composites. Although lauric acid coating of biochar by itself did not have a significant effect on tensile properties, biochar that was first treated with carbon dioxide and then coated with lauric acid showed a 19% increase in tensile strength and a 48% increase in toughness. Gas treatment and lauric acid coating of biochar provide relatively simple processing techniques to improve the stiffness and tensile strength of biochar as rubber composite filler.

**Keywords:** biochar; surface treatment; lauric acid; rubber composite

**Citation:** Peterson, S.C.; Thomas, A.J. Lauric Acid Treatments to Oxidized and Control Biochars and Their Effects on Rubber Composite Tensile Properties. C **2022**, 8, 58. https://doi.org/10.3390/c8040058

Academic Editors: Indra Pulidindi, Pankaj Sharma and Aharon Gedanken

Received: 29 September 2022
Accepted: 27 October 2022
Published: 29 October 2022

**Publisher's Note:** MDPI stays neutral with regard to jurisdictional claims in published maps and institutional affiliations.

**Copyright:** © 2022 by the authors. Licensee MDPI, Basel, Switzerland. This article is an open access article distributed under the terms and conditions of the Creative Commons Attribution (CC BY) license (https://creativecommons.org/licenses/by/4.0/).

## 1. Introduction

Carbon black (CB) has been the dominant filler in the tire industry for over a century [1]. However, there are important reasons why sustainable alternatives should be pursued. CB is a petroleum product, and dependence on foreign fossil fuel sources is challenging since oil prices can be very unstable due to global conflicts and other political reasons [2]. Shrinking CB's footprint in the global marketplace also improves air quality with less pollution [3].

Biochar has been a popular candidate as a sustainable source of carbon from biomass [4]. Most biochar research has been carried out with intended applications being carbon sequestration [5], catalysts [6–8], energy storage [9–11], filtration media [12,13], and sorptive media [14–16], the latter two taking advantage of the (frequently) porous nature of most biochars, although porosity does depend on many factors such as feedstock and processing conditions [17]. However, when using biochar as rubber composite filler, porosity is not as important as other characteristics such as carbon content and particle size. Current literature for using biochar as rubber composite filler is limited but slowly growing. Greenough has reviewed the relevance of biochar as rubber composite filler [18]. Current examples include Xue et al. [19] using rice husk, with its high silica content, as feedstock to make biochar that was ball-milled and used as reinforcing filler. Other sustainable feedstocks studied for biochar use as rubber composite filler include waste lignin [20] and leaf biomass [21].

Suliman and coauthors have studied biochar surface chemistry extensively; both in how feedstock and pyrolysis temperature affect it [22] as well as how oxidation by air has a role in modifying it [23]. Suliman found that for poplar wood as a biochar feedstock, greater quantities of carbonyl and other carboxyl groups were formed on biochar surfaces produced at lower temperatures of 300–400 °C as opposed to higher temperatures ranging up to 600 °C. The formation of these oxygenated functional groups would increase negative charges on the biochar. Carbon dioxide was chosen for comparison with air as a biochar gas treatment. Yi and co-workers studied the influence of $CO_2$ on cellulose biochar properties and found that it enhanced carbon content and increased the C:O ratio [24], and others [25,26] have also observed that gaseous $CO_2$ can react favorably to increase the carbon content in biochar production.

For biochar used as rubber composite filler, modifying the surface chemistry to make the biochar more hydrophobic will improve dispersion in the rubber matrix. Traditionally this has been done using stearic acid [27] (pp. 170–171). Since most biochars have highly charged surfaces [28], adding surfactants creates a 'coated' biochar that will then have an outer surface with many aliphatic long-chain hydrophobic tail groups and thus make the biochar particles themselves more hydrophobic. Navarathna et al. employed this method by saturating biochar with lauric acid (LA) to make it more hydrophobic [29]. Navarathna proposed that the polar carboxylic acid groups in LA are attracted to the oxygenated functional groups on the surface of the biochar, which results in the hydrocarbon chains creating a more hydrophobic outer surface layer on the biochar particles. We hypothesize that this same mechanism will work in dry milling LA with biochar as carried out in this manuscript.

In this work, our goal was to study these two surface chemistry modification methods as a two-step process for biochar, with the first step being gas treatment of the biochar. Using a technique similar to that of Suliman [23], separate samples of poplar biochar were treated with either air or carbon dioxide. In the secondary step, each of these three biochar samples (untreated biochar control, air-treated, or $CO_2$-treated) were then coated with lauric acid using a method similar to that of Navarathna [29], but altered to a dry-milling method more amenable (and scalable) to the rubber compounding industry. Our hypothesis was that LA-coated biochar, being more hydrophobic than uncoated biochar, would have better dispersion in the rubber matrix during compounding and result in more strongly reinforced rubber composites that would show higher tensile strength and/or stiffness than uncoated biochar samples. We were also interested to observe any differences in the gas treatments combined with LA coating, and to determine if the gas treatments acted as favorable or unfavorable pre-treatments for LA coating in terms of final tensile properties of the composites.

## 2. Materials and Methods
### 2.1. Materials Used

Biochar samples made from *Populus tremuloides* (commonly known as poplar) were provided by Green Carbon Nanostructures Corporation (Hartland, WI, USA). Three poplar biochar samples were provided: (1) an unadulterated biochar control, designated "BC control"; (2) a biochar sample that was air-oxidized at 300 °C for 30 min designated "BC air", and (3) a biochar sample treated with carbon dioxide at 300 °C for 30 min designated "BC $CO_2$". An N-339 CB (Vulcan M) was supplied by the Cabot Corporation (Alpharetta, GA, USA). Styrene-butadiene rubber (SBR) provided by Michelin (Greenville, SC, USA) was used as the rubber matrix for composite samples. Lauric acid (98%) was provided by Aldrich Chemical Company (Milwaukee, WI, USA). Precipitated stearic acid-coated calcium carbonate (SA-CC) nanopowder was bought from US Research Nanomaterials, Inc. (Houston, TX, USA) and used as a dry-milling aid.

### 2.2. Chemical and Physical Material Properties

Elemental analysis was done using a PerkinElmer 2400 CHNS/O series II analyzer (Waltham, MA, USA) using cysteine as a standard. Each measurement used approximately 2 mg of biochar and was done in triplicate. Ash content of the biochar samples was determined using a TA Instruments Q2950 thermogravimetric analyzer (New Castle, DE, USA) by heating to 1000 °C at a heating rate of 10 °C per minute in an air atmosphere.

Ash content was determined to be the weight percentage remaining, and oxygen (O) was determined by difference from the original dried sample and the sum of C, H, N, and ash. Densities of the biochar samples were measured in triplicate using a Micromeritics Accupyc II 1340 helium pycnometer (Norcross, GA, USA) using the 10 cm$^3$ sample cup. Infrared spectra were obtained using a Frontier attenuated total reflectance (ATR) Fourier-Transform Infrared Spectrometer (FTIR) (Perkin Elmer, Waltham, MA, USA) fitted with a diamond crystal. The background scan was conducted under ambient atmosphere. Powdered samples were placed directly on the crystal. For each sample, 64 scans were conducted from 650–4000 cm$^{-1}$ at a spectral resolution of 4 cm$^{-1}$. The baseline correction function of the Spectrum software was utilized for all samples.

*2.3. Lauric Acid Coating of Biochar*

The biochar surface was modified by taking 24 g biochar, adding 2.4 g SA-CC and 2.4 g lauric acid, and dry milling this mixture with an SFM-1 (model QM-3SP2) planetary ball mill (MTI Corporation, Richmond, CA, USA). The optimum quantity of lauric acid to use was a matter of investigation and is described in Section 3.2. The milling media used was yttrium-stabilized zirconia spheres (3 mm diameter) supplied by the Inframat Corporation (Manchester, CT, USA). A 50:1 weight ratio of milling media:biochar was used with 500 mL capacity stainless steel jars and lids. Ball milling was done at 500 rpm for 30 min, the samples decelerated to zero rpm and rested for 6 min, then were increased to 500 rpm for 30 min in the opposite direction.

*2.4. Creation and Tensile Testing of Rubber Composite Samples*

Table 1 below contains the recipe for the rubber composite masterbatch used to make rubber composite samples. Amounts listed are in parts per hundred (phr). Each sample was differentiated according to which biochar filler was used; see the following list with their shorthand notation in parentheses: (1) biochar control (BC control); (2) air-oxidized biochar (BC air); (3) carbon dioxide treated biochar (BC CO$_2$); (4) biochar control coated with lauric acid (BC control LA); (5) air-oxidized biochar coated with lauric acid (BC air LA); (6) carbon dioxide treated biochar coated with lauric acid (BC CO$_2$ LA).

**Table 1.** Rubber composite formulation (phr).

| SBR | CB | Filler | MBTBM | SA | ZnO | Sulfur | CBTS |
|---|---|---|---|---|---|---|---|
| 100 | 30 | 12.86 | 0.80 | 2.00 | 3.00 | 2.00 | 1.00 |

SBR: styrene butadiene rubber; CB: carbon black; MBTBM: 2,2′-methylenebis(6-tert-butyl-4-methylphenol); SA: stearic acid; ZnO: zinc oxide; CBTS: N-cyclohexyl-2-benzothiazolesulfenamide.

Tensile properties were measured on an Instron 55R1123C5420 (Instron, Inc., Norwood, MA, USA) and data was processed using Bluehill Universal Software version 4.32. For each rubber composite sample at least 4 replicates were run.

### 3. Results

*3.1. Biochar Characterization*

Physical characteristics of the biochar samples can be seen in Table 2 below. Carbon content in these samples is very high, although not approaching the >99% purity of CB. Density of these biochars is also nearly identical to the density of CB used (1.7–1.9 g/cm$^3$).

Relative purity of these biochars are also shown in the X-ray diffraction spectra seen in Figure 1. Graphitic d-spacing peaks that are typical in CB and other high-carbon containing materials at 24 and 43° 2θ [30] are most prominent in the CB trace, but can also be seen for the biochar samples. No other sharp peaks are seen, meaning there are no appreciable crystalline impurities present in any of the biochar samples.

Table 2. Material properties of carbon black and biochar samples.

| Sample | C (%) | H (%) | N (%) | O (%) [a] | Ash (%) | Density (g/cm³) |
|---|---|---|---|---|---|---|
| CB [b] | >99 | <1 | <1 | <1 | <1 | 1.7–1.9 |
| BC control | 87.22 ± 0.50 | 1.81 ± 0.18 | 0.12 ± 0.04 | 6.52 | 4.33 | 1.68 |
| BC air | 86.42 ± 0.38 | 1.63 ± 0.20 | 0.13 ± 0.02 | 7.93 | 3.89 | 1.69 |
| BC $CO_2$ | 88.67 ± 0.32 | 1.49 ± 0.03 | 0.15 ± 0.03 | 5.96 | 3.73 | 1.69 |

[a] oxygen calculated by difference; [b] data supplied by the manufacturer.

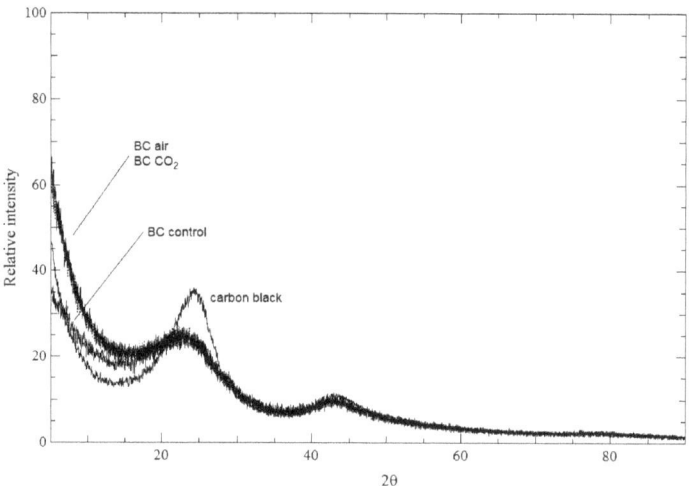

**Figure 1.** X-ray diffraction spectra of CB, biochar control, and gas-treated biochar samples. Graphitic d-spacing peaks at 24° and 43° 2θ are typical of high carbon content materials such as CB and are seen in all four samples.

### 3.2. Determination of Lauric Acid Coating Concentration

To determine the appropriate amount of LA to use to coat biochar particles, several different concentrations (5, 10, and 20% relative to biochar weight) of LA were milled with a control biochar. The 5, 10, and 20% LA-coated control biochar samples were then molded into rubber composite samples and their tensile properties were measured. Results can be seen in Table 3.

**Table 3.** Tensile properties of biochar as a function of LA concentration.

| LA Concentration (%) | Tensile Strength (MPa) | Elongation (%) | Toughness (MPa) |
|---|---|---|---|
| 0 | 19.8 ± 0.4 | 526 ± 14 | 49.1 ± 2.7 |
| 5 | 19.2 ± 0.5 | 527 ± 5 | 46.3 ± 1.8 |
| 10 | 20.5 ± 1.1 | 552 ± 20 | 53.2 ± 3.7 |
| 20 | 19.3 ± 0.8 | 513 ± 10 | 45.6 ± 0.9 |

Tensile results over the LA range tested for these rubber composites were similar, but since the 10% LA-coated composite had a slightly higher tensile strength and increased toughness, this concentration was selected to coat all the biochar samples.

Confirmation of LA coating on biochar samples can be seen by the presence of the characteristic LA C-H stretching vibrations at 2915 and 2847 cm$^{-1}$ [31] in Figure 2. These biochars were then used to create rubber composite samples for tensile testing.

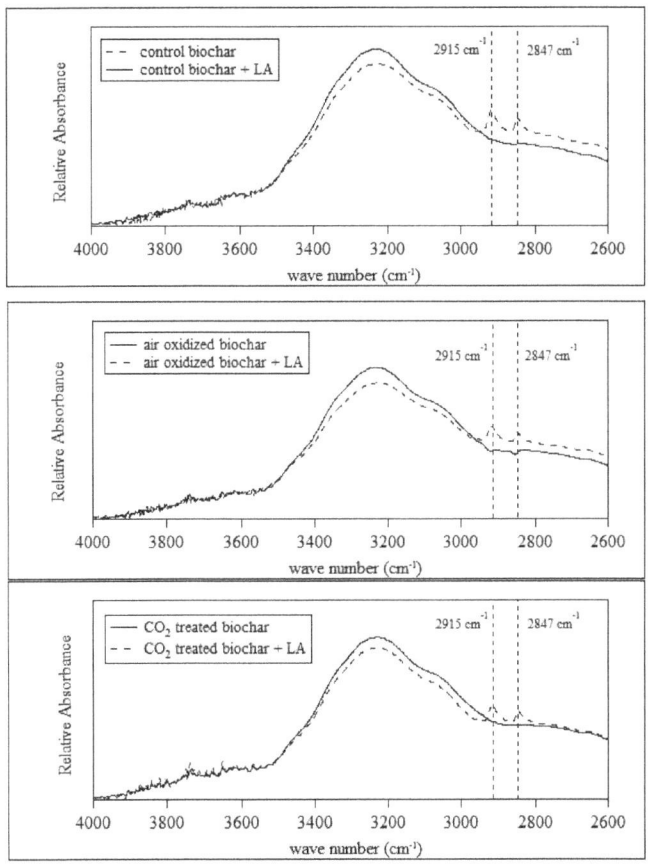

**Figure 2.** FTIR spectra for LA-coated vs. uncoated biochar samples showing the presence of LA by the peaks at 2847 and 2915 cm$^{-1}$.

*3.3. Tensile Property Results*

Tensile results for both uncoated and 10% LA-coated biochar samples are shown in Table 4. The effect of gas treatments to the biochar with air and $CO_2$ can be seen in the first three rows of Table 4. Tensile strength remains similar with all three samples, however both gas treatments reduce elongation relative to the control.

**Table 4.** Tensile properties of uncoated and 10% LA-coated biochar samples.

| Sample | n | Tensile Strength (MPa) | Elongation (%) | Toughness (MPa) |
| --- | --- | --- | --- | --- |
| BC control | 8 | 20.1 ± 2.1 | 501 ± 42 | 43.1 ± 8.5 |
| BC air | 5 | 21.2 ± 1.0 | 488 ± 17 | 44.8 ± 4.0 |
| BC $CO_2$ | 4 | 18.6 ± 1.1 | 447 ± 20 | 34.9 ± 3.9 |
| BC control LA | 5 | 20.7 ± 0.9 | 532 ± 18 | 47.9 ± 3.7 |
| BC air LA | 4 | 20.8 ± 1.3 | 504 ± 25 | 45.4 ± 5.7 |
| BC $CO_2$ LA | 5 | 22.1 ± 0.6 | 536 ± 12 | 51.6 ± 2.7 |

$n$ = number of replicate runs.

By examining the stress–strain curves for these three samples (Figure 3), the higher stiffness of the gas-treated samples is evident in the steeper slopes compared to that of the control. It should be noted that in all stress–strain plots presented in this work, the representative curve that was closest to the average tensile strength and elongation for each sample was selected and shown for clarity.

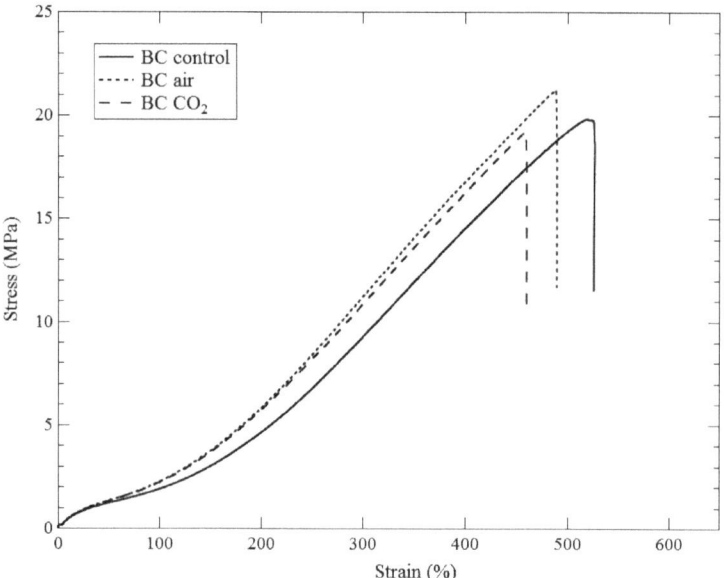

**Figure 3.** Stress vs. strain curve for the three uncoated biochar samples, illustrating the effect of air and $CO_2$ treatment during biochar processing.

To show the effect of LA coating, Figure 4 shows each of the stress–strain curves for the BC control, BC air, and BC $CO_2$ samples with and without LA coating. For the control biochar, LA coating showed no appreciable change in the tensile strength or elongation as the stress–strain curves were nearly identical. For the air-treated biochar sample, tensile strength remained unchanged and there was a slight increase in stiffness relative to the uncoated sample. The effect of LA coating on the $CO_2$ treated biochar sample showed that stiffness was essentially unchanged, but the tensile strength increased nearly 20%.

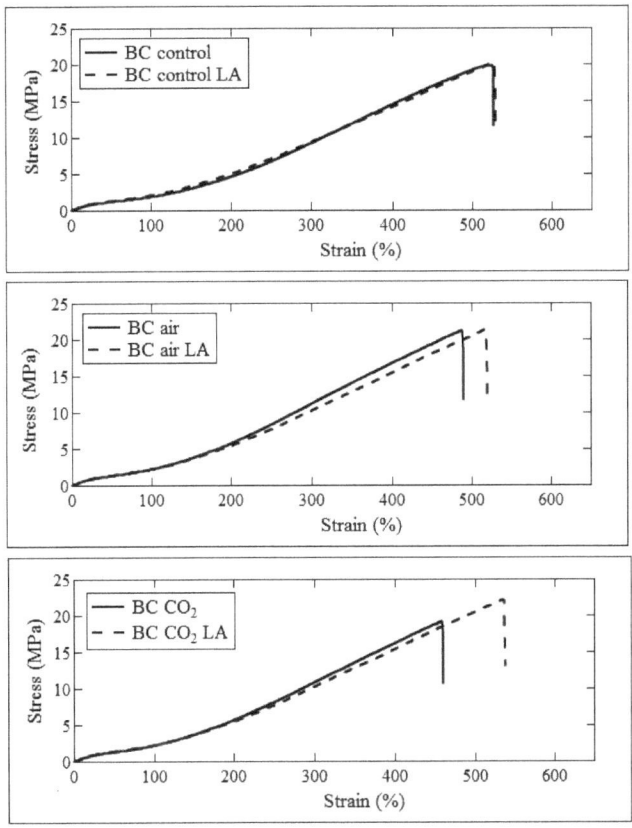

**Figure 4.** Stress–strain curves of LA-coated vs. uncoated rubber composite samples for the control (**top**), air-treated (**middle**), and $CO_2$-treated (**bottom**) biochar samples. For each plot the single representative curve closest to the average tensile strength and elongation was chosen for clarity.

## 4. Conclusions

For this work we wanted to observe two separate surface chemistry modifications to biochar; a gas treatment step followed by the addition of LA. Both treatments are amenable to the rubber processing industry as they only involve gas treatment to the biochar at controlled temperature and dry-milling, two processes that are both easily scalable. Gas treatment of the biochars with air and $CO_2$ at 300 °C did not appreciably affect the tensile strength of the resulting filled rubber composites, but did increase their stiffness relative to the untreated biochar control.

Biochar control coated with LA appeared to have no effect on the tensile properties of the resulting filled rubber composite. For the air-treated biochar sample, coating with LA did not appreciably change the tensile strength, but did slightly increase the elongation of the composite, making it slightly softer. The most notable difference in tensile properties occurred when the $CO_2$-treated biochar was then LA-coated; composites made from this biochar sample increased their tensile strength by 19%, and their toughness by 48%. This suggests that the $CO_2$ gas treatment helps condition the biochar so that LA coating is more effective. $CO_2$ treatment of biochar increased its carbon content and reduced its oxygen content, reducing the number of charged, oxygen-containing functional groups on the biochar and making it more hydrophobic. Further investigation to understand the relative roles of $CO_2$ treatment in conjunction with LA coating is needed to help optimize and improve biochar reinforcement of rubber composites.

**Author Contributions:** Conceptualization, S.C.P.; methodology, S.C.P.; investigation, S.C.P. and A.J.T.; resources, A.J.T.; writing—original draft preparation, S.C.P.; writing—review and editing, S.C.P. and A.J.T. All authors have read and agreed to the published version of the manuscript.

**Funding:** This work was supported by the U.S. Department of Agriculture, Agricultural Research Service. Mention of trade names or commercial products in this publication is solely for the purpose of providing specific information and does not imply recommendation or endorsement by the U.S. Department of Agriculture. USDA is an equal opportunity provider and employer.

**Acknowledgments:** The authors would like to thank Jason Adkins for collecting ash content data and Kelly Utt for performing FTIR experiments.

**Conflicts of Interest:** The authors declare no conflict of interest.

# References

1. Carbon Black in Tires. Available online: https://www.orioncarbons.com/tire (accessed on 30 September 2022).
2. Lee, C.-C.; Olasehinde-Williams, G.; Akadiri, S.S. Are geopolitical threats powerful enough to predict global oil price volatility? *Environ. Sci. Pollut. Res.* **2021**, *28*, 28720–28731. [CrossRef] [PubMed]
3. Ragothaman, A.; Anderson, W.A. Air Quality Impacts of Petroleum Refining and Petrochemical Industries. *Environments* **2017**, *4*, 66. [CrossRef]
4. Liu, W.-J.; Jiang, H.; Yu, H.-Q. Development of Biochar-Based Functional Materials: Toward a Sustainable Platform Carbon Material. *Chem. Rev* **2015**, *115*, 12251–12285. [CrossRef] [PubMed]
5. Gross, A.; Bromm, T.; Glaser, B. Soil Organic Carbon Sequestration after Biochar Application: A Global Meta-Analysis. *Agronomy* **2021**, *11*, 2474. [CrossRef]
6. Cheng, F.; Li, X. Preparation and Application of Biochar-Based Catalysts for Biofuel Production. *Catalysts* **2018**, *8*, 346. [CrossRef]
7. Wang, R.-Z.; Huang, D.-L.; Liu, Y.-G.; Zhang, C.; Lai, C.; Wang, X.; Zeng, G.-M.; Gong, X.-M.; Duan, A.; Zhang, Q.; et al. Recent advances in biochar-based catalysts: Properties, applications and mechanisms for pollution remediation. *Chem. Eng. J.* **2019**, *371*, 380–403. [CrossRef]
8. Do Minh, T.; Song, J.; Deb, A.; Cha, L.; Srivastava, V.; Sillanpää, M. Biochar based catalysts for the abatement of emerging pollutants: A review. *Chem. Eng. J.* **2020**, *394*, 124856. [CrossRef]
9. Liu, W.-J.; Jiang, H.; Yu, H.-Q. Emerging applications of biochar-based materials for energy storage and conversion. *Energy Environ. Sci.* **2019**, *12*, 1751–1779. [CrossRef]
10. Senthil, C.; Lee, C.W. Biomass-derived biochar materials as sustainable energy sources for electrochemical energy storage devices. *Renew. Sustain. Energy Rev.* **2021**, *137*, 110464. [CrossRef]
11. Ehsani, A.; Parsimehr, H. Electrochemical energy storage electrodes from fruit biochar. *Adv. Colloid Interface Sci.* **2020**, *284*, 102263. [CrossRef] [PubMed]
12. Xiang, W.; Zhang, X.; Chen, J.; Zou, W.; He, F.; Hu, X.; Tsang, D.C.W.; Ok, Y.S.; Gao, B. Biochar technology in wastewater treatment: A critical review. *Chemosphere* **2020**, *252*, 126539. [CrossRef] [PubMed]
13. Palansooriya, K.N.; Yang, Y.; Tsang, Y.F.; Sarkar, B.; Hou, D.; Cao, X.; Meers, E.; Rinklebe, J.; Kim, K.-H.; Ok, Y.S. Occurrence of contaminants in drinking water sources and the potential of biochar for water quality improvement: A review. *Crit. Rev. Env. Sci. Technol.* **2020**, *50*, 549–611. [CrossRef]
14. Gwenzi, W.; Chaukura, N.; Wenga, T.; Mtisi, M. Biochars as media for air pollution control systems: Contaminant removal, applications and future research directions. *Sci. Total Environ.* **2021**, *753*, 142249. [CrossRef] [PubMed]
15. Peiris, C.; Nawalage, S.; Wewalwela, J.J.; Gunatilake, S.R.; Vithanage, M. Biochar based sorptive remediation of steroidal estrogen contaminated aqueous systems: A critical review. *Environ. Res.* **2020**, *191*, 110183. [CrossRef] [PubMed]
16. Ghosh, D.; Maiti, S.K. Can biochar reclaim coal mine spoil? *J. Environ. Manag.* **2020**, *272*, 111097. [CrossRef] [PubMed]
17. Weber, K.; Quicker, P. Properties of biochar. *Fuel* **2018**, *217*, 240–261. [CrossRef]
18. Greenough, S.; Dumont, M.-J.; Prasher, S. The physicochemical properties of biochar and its applicability as a filler in rubber composites: A review. *Mater. Today Commun.* **2021**, *29*, 102912. [CrossRef]
19. Xue, B.; Wang, X.; Sui, J.; Xu, D.; Zhu, Y.; Liu, X. A facile ball milling method to produce sustainable pyrolytic rice husk bio-filler for reinforcement of rubber mechanical property. *Ind. Crop. Prod.* **2019**, *141*, 111791. [CrossRef]
20. Jiang, C.; Bo, J.; Xiao, X.; Zhang, S.; Wang, Z.; Yan, G.; Wu, Y.; Wong, C.; He, H. Converting waste lignin into nano-biochar as a renewable substitute of carbon black for reinforcing styrene-butadiene rubber. *Waste Manag.* **2020**, *102*, 732–742. [CrossRef]
21. Lay, M.; Rusli, A.; Abdullah, M.K.; Abdul Hamid, Z.A.; Shuib, R.K. Converting dead leaf biomass into activated carbon as a potential replacement for carbon black filler in rubber composites. *Compos. Part B Eng.* **2020**, *201*, 108366. [CrossRef]
22. Suliman, W.; Harsh, J.B.; Abu-Lail, N.I.; Fortuna, A.-M.; Dallmeyer, I.; Garcia-Perez, M. Influence of feedstock source and pyrolysis temperature on biochar bulk and surface properties. *Biomass Bioenerg.* **2016**, *84*, 37–48. [CrossRef]

23. Suliman, W.; Harsh, J.B.; Abu-Lail, N.I.; Fortuna, A.-M.; Dallmeyer, I.; Garcia-Perez, M. Modification of biochar surface by air oxidation: Role of pyrolysis temperature. *Biomass Bioenerg.* **2016**, *85*, 1–11. [CrossRef]
24. Yi, Z.; Li, C.; Li, Q.; Zhang, L.; Zhang, S.; Wang, S.; Qin, L.; Hu, X. Influence of CO2 atmosphere on property of biochar from pyrolysis of cellulose. *J. Environ. Chem. Eng.* **2022**, *10*, 107339. [CrossRef]
25. Shen, Y.; Ma, D.; Ge, X. $CO_2$-looping in biomass pyrolysis or gasification. *Sustain. Energy Fuels* **2017**, *1*, 1700–1729. [CrossRef]
26. Jung, S.-H.; Kim, J.-S. Production of biochars by intermediate pyrolysis and activated carbons from oak by three activation methods using $CO_2$. *J. Anal. Appl. Pyrol.* **2014**, *107*, 116–122. [CrossRef]
27. Rothon, R.N. *Rapra Technology Limited. Particulate-Filled Polymer Composites*, 2nd ed.; Rapra Technology: Shrewsbury, UK, 2003; 544p.
28. Mukherjee, A.; Zimmerman, A.R.; Harris, W. Surface chemistry variations among a series of laboratory-produced biochars. *Geoderma* **2011**, *163*, 247–255. [CrossRef]
29. Navarathna, C.M.; Bombuwala Dewage, N.; Keeton, C.; Pennisson, J.; Henderson, R.; Lashley, B.; Zhang, X.; Hassan, E.B.; Perez, F.; Mohan, D.; et al. Biochar Adsorbents with Enhanced Hydrophobicity for Oil Spill Removal. *ACS Appl. Mater. Interfaces* **2020**, *12*, 9248–9260. [CrossRef] [PubMed]
30. Darmstadt, H.; Roy, C.; Kaliaguine, S.; Xu, G.; Auger, M.; Tuel, A.; Ramaswamy, V. Solid state 13C-NMR spectroscopy and XRD studies of commercial and pyrolytic carbon blacks. *Carbon* **2000**, *38*, 1279–1287. [CrossRef]
31. Kong, W.; Fu, X.; Yuan, Y.; Liu, Z.; Lei, J. Preparation and thermal properties of crosslinked polyurethane/lauric acid composites as novel form stable phase change materials with a low degree of supercooling. *RSC Adv.* **2017**, *7*, 29554–29562. [CrossRef]

MDPI
St. Alban-Anlage 66
4052 Basel
Switzerland
Tel. +41 61 683 77 34
Fax +41 61 302 89 18
www.mdpi.com

C Editorial Office
E-mail: carbon@mdpi.com
www.mdpi.com/journal/carbon

www.ingramcontent.com/pod-product-compliance
Lightning Source LLC
LaVergne TN
LVHW070627100526
838202LV00012B/748